Compound Polytopes

polygons, tilings, polyhedra...

Patrick Taylor

POLYSTAR PRESS

ISBN 978 1 907154 62 1

Compound Polytopes

Published by Polystar Press

62 Angel Street
Hadleigh
Suffolk
IP7 5EY
(01473) 824896
polystar@btinternet.com

ISBN 978 1 907154 62 1

Typeset by nattygrafix

Printed by
R Booth Ltd, The Praze, Penryn

CONTENTS

For Rosehay and all who dwelt there

(1955-2018)

INTRODUCTION

This book started out as a re-presentation of material from the author's previous works, the various volumes of his *'The Complete? Polyhedra'*. Although the selections made do not offer a complete survey of that subject, they concentrate on the new discoveries within each volume that follow on from taking a different approach to defining a polygon.

First published in 1997 *'The Complete? Polygon'* proposed an alternative notation and was followed by a series of publications that allowed the higher geometries of tilings and polyhedra (assemblies of polygons) and then closepacks and hypersolids (assemblies of polyhedra) to be greatly enriched by allowing the inclusion of a wider vocabulary of polygons than the current geometer's view allows.

This work has not been widely accepted because of its being at variance with the current notation, but it really needs to be looked at in the true spirit of mathematics: 'if we do it this way, then what follows?'.

Within mathematics generally we might choose to use the 'Reverse Notation' or count in base 6, but neither of these choices would change the underlying behaviour of the things we study, they would simply serve to allow us to see more clearly what we are studying and thus obtain a better understanding. Newton and Leibniz had different takes on the calculus, and both were right!

Indeed going back further Euclid's original world is now seen as flat in the light of other geometries that dwell on spherical or hyperbolic spaces. These other worlds were brought about by the simple act of dropping the parallel postulate, but leave Euclid's flat world intact within a wider spectrum of spaces with positive or negative curvature. Here some amazing geometry can be explored and in a similar way it is hoped that this slim volume will open up further unexplored vistas.

So here we go, we are going to do it this way and what follows is enough to fill these pages with a remarkable diversity of material in two, three and indeed four dimensions. We will be exploring many new forms in increasing numbers of dimensions that all very much exist as parallels to the accepted versions of tilings, polyhedra, closepacks and hypersolids, where symmetry families comprise truncation series of related forms.

Whilst finishing off the book it became apparent that much of the material, like that of the generally accepted regular tilings, polyhedra, closepacks and hypersolids, can be seen to fit within a series of standard polytopes at different dimensional levels.

The standard formulae for the regular simplex, measure and honeycomb polytopes can be altered by a simple substitution, to produce formulae for these new 'compound polytopes'. In the simplest terms this substitution can be seen as nothing more than the stellation or inscription of the regular forms, so obvious and yet apparently unappreciated before now.

So basically what we have here is the adoption of an alternative notation for the polygons, which is inclusive of the compounds and provides rich rewards with a series of new compound geometries at higher dimensions as illustrated herein. Surely if Mathematics is about finding patterns, then rather than being incorrect, the alternative notation as proposed here is in fact a useful tool for discovery.

If we take the view that the Mathematics is out there anyway then the alternative notation proposed here provides a key to some interesting but thus far uncharted corners of geometry. If we think Mathematics is purely a product of the mind and that such new areas are actually invented, then welcome to my world.

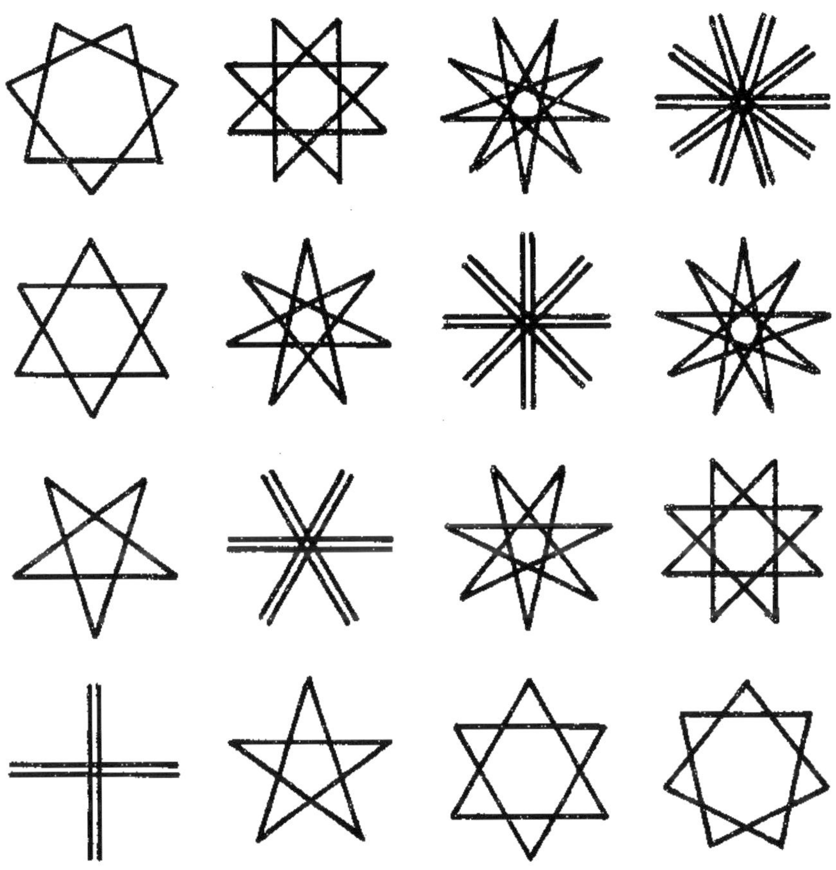

ARE YOUR POLYGONS THE SAME AS MY POLYGONS?

Not necessarily, because it depends on how they are defined. The current view adopted by geometers favours a combinatorial approach involving prime numbers, which ties in with number theory. It defines polygons in such a way that those that can be seen as compounds of simpler polygons are removed from consideration as if they were an embarrassment. This leads at higher levels to a similar simplification amongst the polyhedra, where compounds are also excluded and the full richness of possible forms is not properly appreciated.

Taking this currently held approach we can very simply define a general polygon {n/d} in terms of n equally spaced points around a circle, from <u>one</u> of which an edge is drawn to the dth point around the circle, from where the same operation is repeated until we have n edges and end up at the original starting point. This method has analogies with modular arithmetic.

When d=1 we get the general regular polygon {n/1}, constructed from n points, each connected to the adjacent one by an edge. No problems here.

When d=2 we have to connect with an edge to the next but one point around the circle. This works well when n is odd, as each original point is visited just once, on either the first or second circuit of the circle. However when n is even, after n/2 edges are in place we find ourselves back at the starting point and repeating exactly the same moves for a second circuit. This leaves us with effectively a double coincident polygon and n/2 of our original vertices totally unvisited, which seems somehow unsatisfactory.

Moving up to d=3 we find that when n is a multiple of 3, we get a treble coincident {n/3} polygon with n edges utilising only n/3 vertices, leaving 2n/3 of them unvisited.

One bonus of this particular way of looking at polygons is that the truncation of {n/d} is always {2n/d}, but to get this we have excluded certain polygonal forms, the compounds, from our vocabulary. This is not aesthetically pleasing, nor worthwhile, as many interesting forms are unnecessarily omitted, as shown by their absence in Figure 1 opposite.

This book seeks to address this problem, putting forward an alternative more inclusive way of notating the polygons which allows the inclusion of the missing compound forms. To do this we need two separate rules for truncation, which simply depend on whether d is odd or even, sometimes held up as a reason for not using this alternative method of notation.

We will shortly reveal that to include the missing compound polygons, the currently accepted system also needs two different rules for truncation which are not so easy to differentiate, something which cannot be avoided.

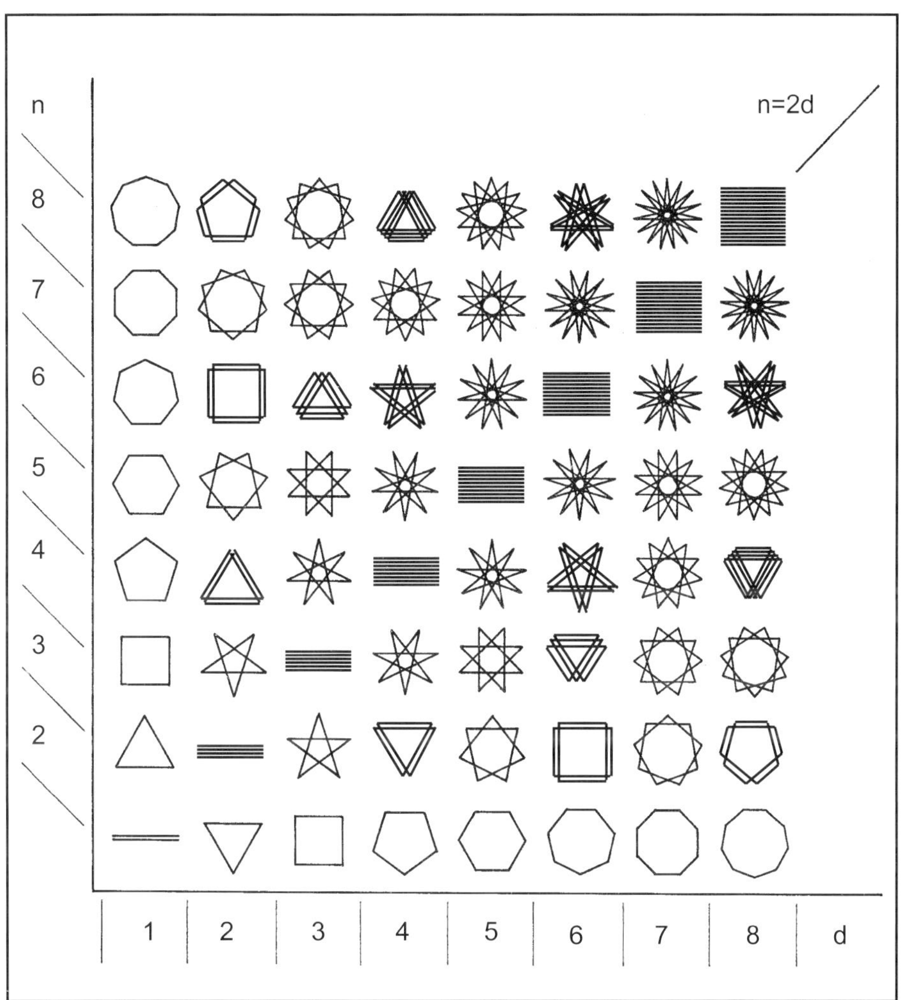

Figure 1: Polygons {n/d} according to current notation

ALTERNATIVE NOTATION

First published in 1997 the present author's *'The Complete? Polygon'* proposed an alternative notation and was followed by a series of publications that allowed the higher geometries of tilings and polyhedra (assemblies of polygons) and then closepacks and hypersolids (assemblies of polyhedra) to be greatly enriched by allowing the inclusion of a wider vocabulary of polygons than the current geometer's view allows.

This work has not been generally accepted because of its variance with the current notation, but it really needs to be looked at in the true spirit of mathematics: 'if we do it this way, then what follows?'.

It must therefore be pointed out before we go any further that it is only the notation that we are varying, the polygons themselves still behave in exactly the same way whatever they are called, especially when truncated. There is no right or wrong way here.

The alternative notation proposed constructs polygons in a very similar manner to the accepted way with one small difference. We simply define a general polygon {n/d} in terms of n equally spaced points around a circle, from <u>each</u> of which an edge is drawn to the dth point around the circle.

Thus the general regular polygon {n/1} is constructed from n points, each connected to the adjacent one by an edge. Similarly the star polygon {n/2} is constructed from n points, each connected to the next but one by an edge.

This method is inclusive and uses all of the vertices proposed without any redundancies. Also in this method there is not any special vertex from which one starts as all the vertices are used.

As n increases the possibilities for different star polygons {n/d} increase and we can find {n/3}, {n/4} and so on, as long as we keep d<n/2.

If we allow d>n/2, we obtain the inverse polygons, basically repeats of the regular and star polygons, but which have a reversed sense of direction. Every polygon x = {n/d} thus has an inverse x' = {n/(n-d)} and the pair x and x' are known as conjugates.

In the special case where d=n/2 (only possible when n is even), we obtain a cross polygon consisting of d pairs of edges back to back, in effect a compound of d digons. These are always self-inverse since d=n−d.

Figure 2 opposite shows these polygons in terms of this more inclusive alternative notation, arrayed in a pleasing symmetrical way, with the compound ones not immediately standing out. We are not doing number theory here!

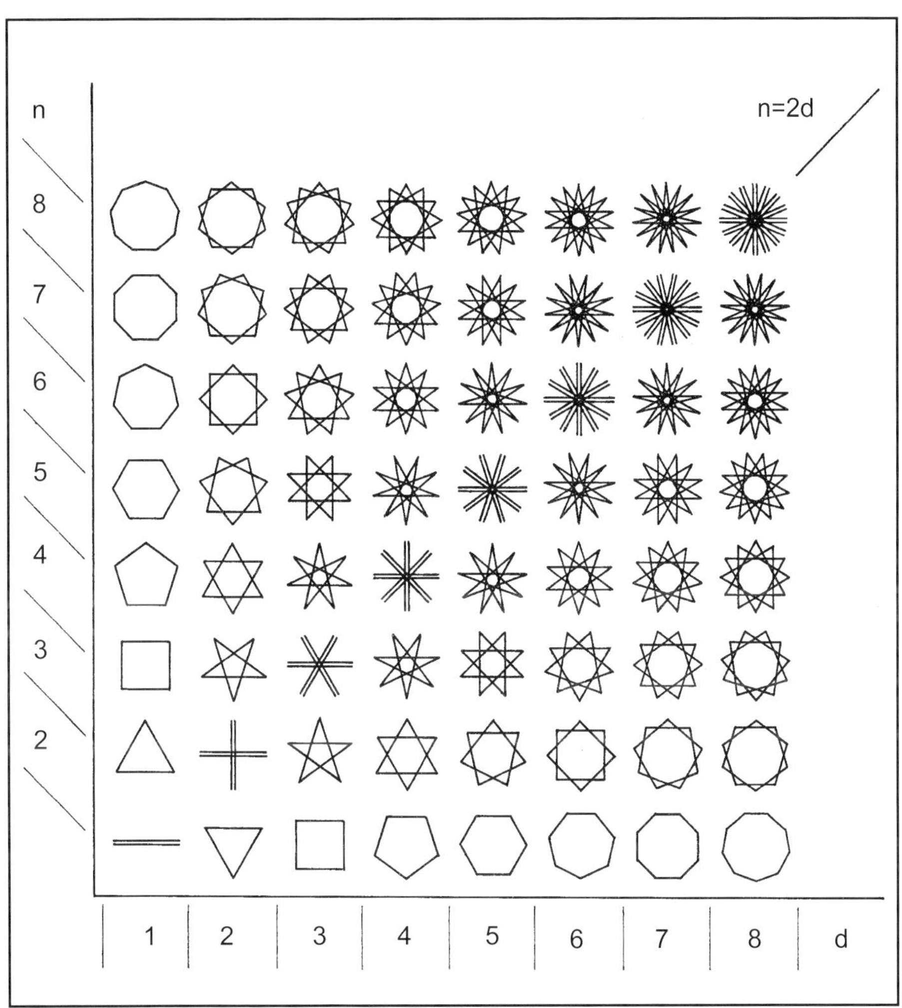

Figure 2: Polygons {n/d} according to alternative notation

INSCRIBED POLYGONS

Starting with the regular polygons {n/1}, which are the same in both notation systems, we can find within each a series of 'star' polygons by inscribing new edges that link across the internal space to more distant vertices. This is essentially the same process as that used to define the polygons in the alternative notation, where each vertex is joined by a new edge to the dth point around the circle of vertices.

Sticking firstly to examples where d<n/2 we obtain a series of internal polygons with increasing sharpness of vertex and decreasing internal angle {n/2}, {n/3}, {n/4} etc. When n is even and d=n/2, the vertices are at their sharpest with zero internal angle, the polygons created comprising d crossed digons forming what we will call a cross polygon. Beyond here the vertices decrease in sharpness again and have effectively increasing negative internal angles until the inverse polygon {n/n-1} is reached.

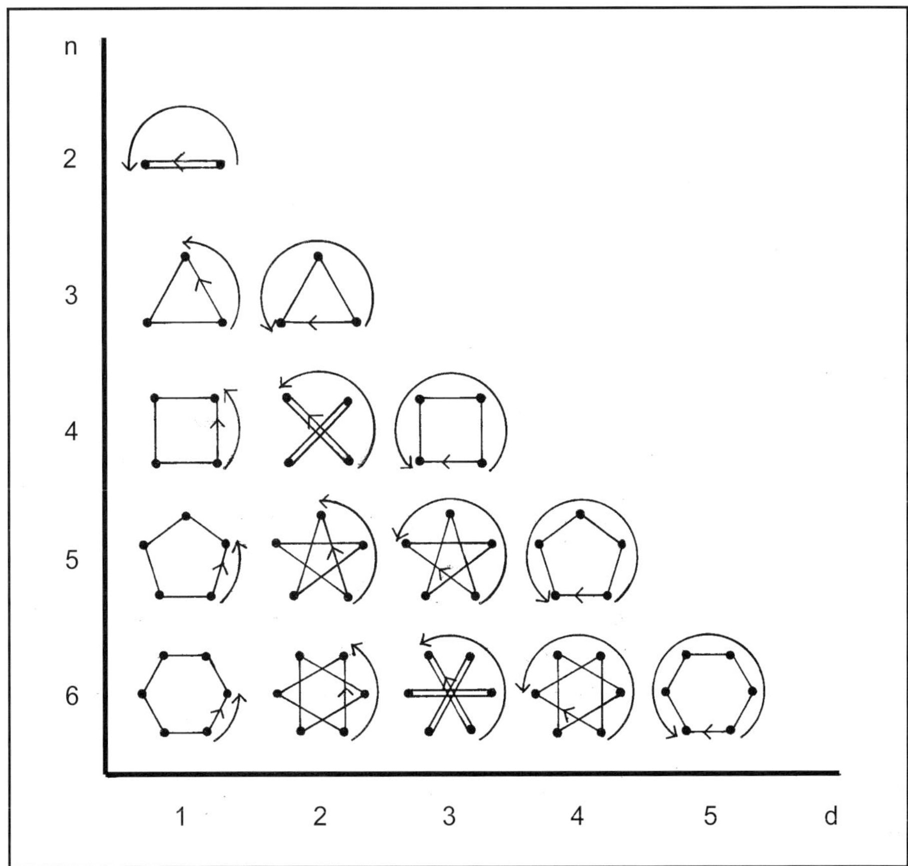

Figure 3: Inscriptions of the regular polygons, n = 2, 3, 4, 5 & 6

Exactly the same series of star polygons can also be achieved by stellation. Here we take an edge of the regular polygon {n/1} and extend it along with the dth edge around until they cross at a new vertex position. In the cases where d=n/2 the crossing is effectively at infinity, but this can be normalised by adopting a set standard edge length to match any others they are being assembled with, so that these polygons then have no thickness to their arms and no actual area.

These ways of looking at the creation of polygons tie in with our alternative definition and are the same for all, without any inbuilt prejudice against the compound polygons. These last will hopefully prove themselves worthy of further study when we include them in various new tiling and polyhedra families, where they perform exactly the same functions as the accepted non -compounds generally used.

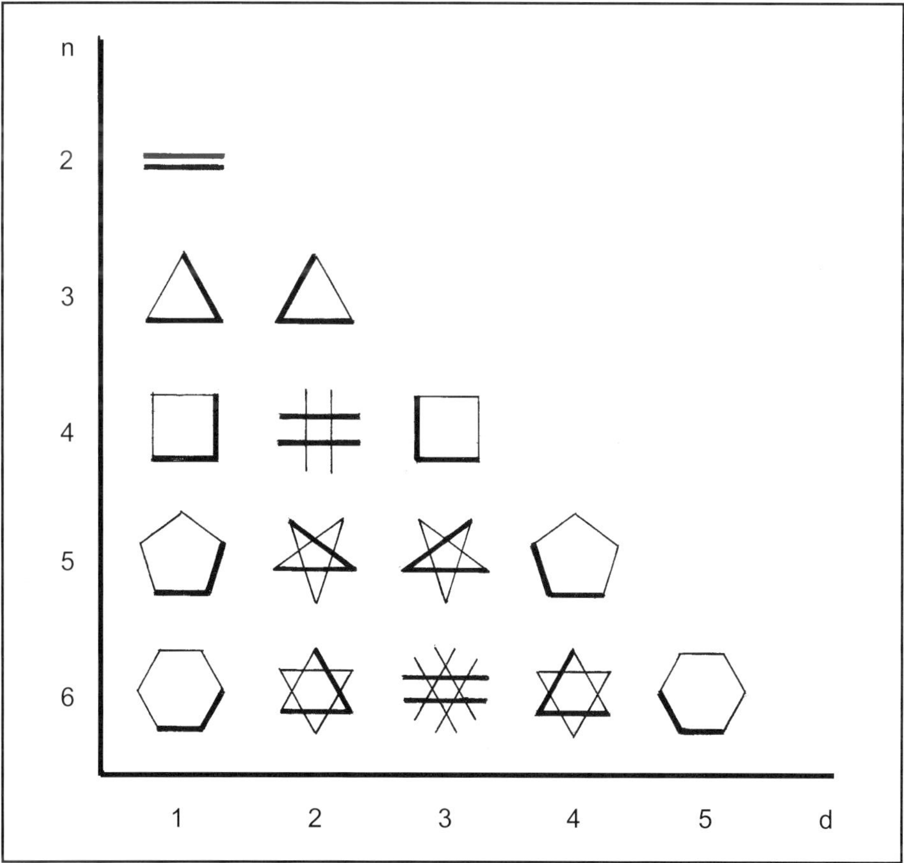

Figure 4: Stellations of the regular polygons, n = 2, 3, 4, 5 & 6

TRUNCATION RULES

Compared to the {5/1} or {7/1} series where n is prime, the only difference with the {6/1} series of inscriptions or stellations is that they are effectively compound, because 6 is not co-prime with 2, 3 or indeed 4. The currently adopted approach deals with these compounds in a somewhat disjointed way, calling the first inscription or stellation of the {6/1} polygon series 2{3/1}, the second 3{2/1} and the third 2{3/2}.

The present author would simply classify these as {6/2}, {6/3} and {6/4} respectively, analogous to the similar series for {5/1} and {7/1}. He would then predict the truncations of these to be (in his new notation) ^{2}{6/1}, {12/3} and ^{2}{6/2}. The two rules used here in the alternative version are that if d is odd then t{n/d} = {2n/d}, but if d is even then t{n/d} = ^{2}{n/(d/2)}, the new symbol ^{2} indicative of a double coincident polygon.

Now in the current view the notation {6/2} is reserved for two coincident triangles, {6/3} for three coincident digons and {6/4} for two coincident inverse triangles (see Figure 1). In that view their truncations are quite correctly {12/2}, {12/3} and {12/4}, which are two coincident hexagons, three coincident squares and four coincident triangles respectively. This requires only the seemingly simpler truncation rule in the currently adopted notation that the truncated polygon t{n/d} = {2n/d}, but this does not deal with how we might truncate the compounds.

In the alternative version we will be using here, {6/2} is still two triangles but they are out of phase forming a compound 'Star of David'. The truncation of this is again two coincident hexagons, so that using the current notation we get t{2(3/1)} = {12/2} (a).

In the alternative version {6/3} is three out of phase digons visiting all six vertices and its truncation is {12/3}, three out of phase squares visiting all twelve vertices. In the current view this is notated as t{3(2/1)} = {3(4/1)} (b).

Finally in the alternative version {6/4} is an inverse Star of David formed of two out of phase inverse triangles. It truncates to a double Star of David, notated in the current view as t{2(3/2)} = {2(6/2)} (c).

We thus find that if we include compounds in the current notation there are indeed two rules for their truncation. Put simply the truncation of the general polygon t{m(n/d)} = {m(2n/d)} when (b) m is odd, or (c) if m and d are both even. However there has to be a second rule that t{m(n/d)} = {m/2(4n/2d)} when (a) m is even and d odd, which is not so simple, but unfortunately that is the way it works with this notation if the compounds are to be included.

The current view thus only really deals with the non-compound polygons and universally employs the first of these truncation rules with m set at 1 (odd). It does not deal adequately with the case of compound polygons where two separate rules are in fact needed for truncation, and the application of those rules is far from straightforward.

Hopefully we can now see the possibilities here of a more encompassing alternative classification of the polygons, naturally linked to the process of inscription or stellation and employing a simpler rule for truncation that acknowledges the arising of double polygons as the result of truncation when d is even.

From here onwards we will work solely with this new alternative notation with a view to exploring the world thrown up by including the compound polygons as equal partners alongside the non-compound polygons already accepted within the current notation.

The remainder of this book will persevere with this new notation because although it involves two distinct rules for truncation, it does provide a consistent view of the polygons. Whilst others may choose to define their polygons differently, hopefully the rewards of applying this new view will become apparent as we proceed to open up new areas of geometry not previously taken seriously.

TRUNCATING POLYGONS

In general the truncation of a polygon {n/d}, which has n edges, involves the introduction of a further n edges of equal length to produce a polygon with 2n edges. This is most easily visualised for a regular polygon as the cutting off of corners, turning {n/1} into {2n/1}. It can also be seen as the halving of the external angles of the polygon, so that the angle that was turned at one vertex, is now shared between two vertices, where a 'truncating' edge is introduced between two 'truncated' edges.

When d is odd, this general principle can be applied to all polygons {n/d} to give a truncation {2n/d} for which the external angle will change from 360d/n° in the original to 180d/n° in the truncation. In the case when d is even, the external angle is similarly halved, but each 'truncating' edge coincides with a 'truncated' one, so that we obtain a double polygon which we will notate here as ^{2}{n/(d/2)}.

The polygons produced by the truncation of regular and star polygons {n/d}, when d<n/2, can easily be interpreted in terms of this cutting off of corners. However with the inverse polygons, when d>n/2, although the polygons produced follow the same numerical rules as during normal truncation, the process becomes less clear and is termed quasitruncation.

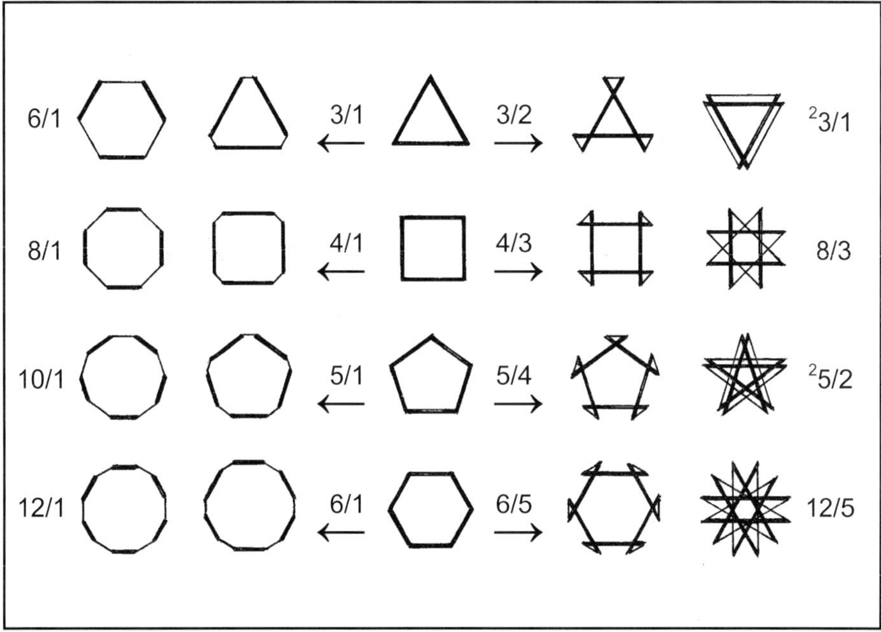

Figure 5: Truncation and quasitruncation of regular polygons, n = 3, 4, 5 & 6

An inverse polygon with d>n/2 has an external angle defined by the same rule, i.e. 360d/n° in the original which still becomes 180d/n° in the truncation. The difference is that since the edges go in the opposite direction around the original polygon, this external angle starts off greater than 180° and when halved will remain greater than 90°.

A simple way to visualise what is happening during quasitruncation is to imagine normal truncation to involve the pulling apart of the original edges to make room for the new truncating edges, whilst quasitruncation involves the pushing together of these same edges, until they cross past the centre of the polygon to make room for the new truncating edges.

Figures 5 and 6 below show the results of truncating and quasitruncating the polygons {n/d}, all those with n between 3 and 6. These relatively simple results are all that we will need to pursue the creation of the new families of compound tilings and polyhedra that result from introducing the previously disregarded compound polygons into the mix.

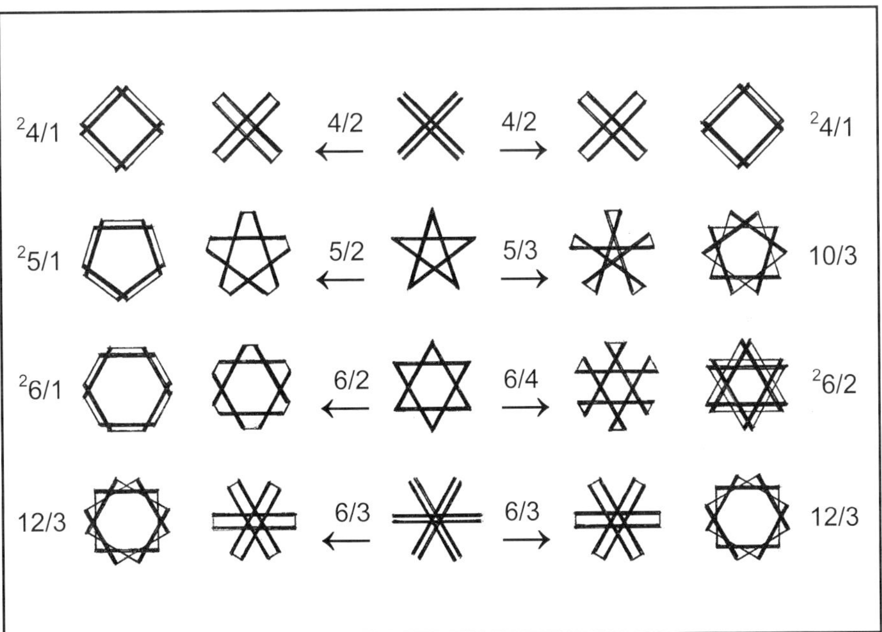

Figure 6: Truncation & quasitruncation of cross & star polygons, n = 4, 5 & 6

REFERENCES

Grünbaum, B. 2003 *Are Your Polyhedra the Same as My Polyhedra?* In: Aronov, Basu, Pach & Sharir (eds) Discrete and Computational Geometry. Algorithms and Combinatorics, vol 25 Springer, Berlin, Heidelberg

Johnston, J.H. 1937 *The Reverse Notation* Blackie & Son Ltd

Taylor, P. 1997 *The Complete? Polygon* Nattygrafix

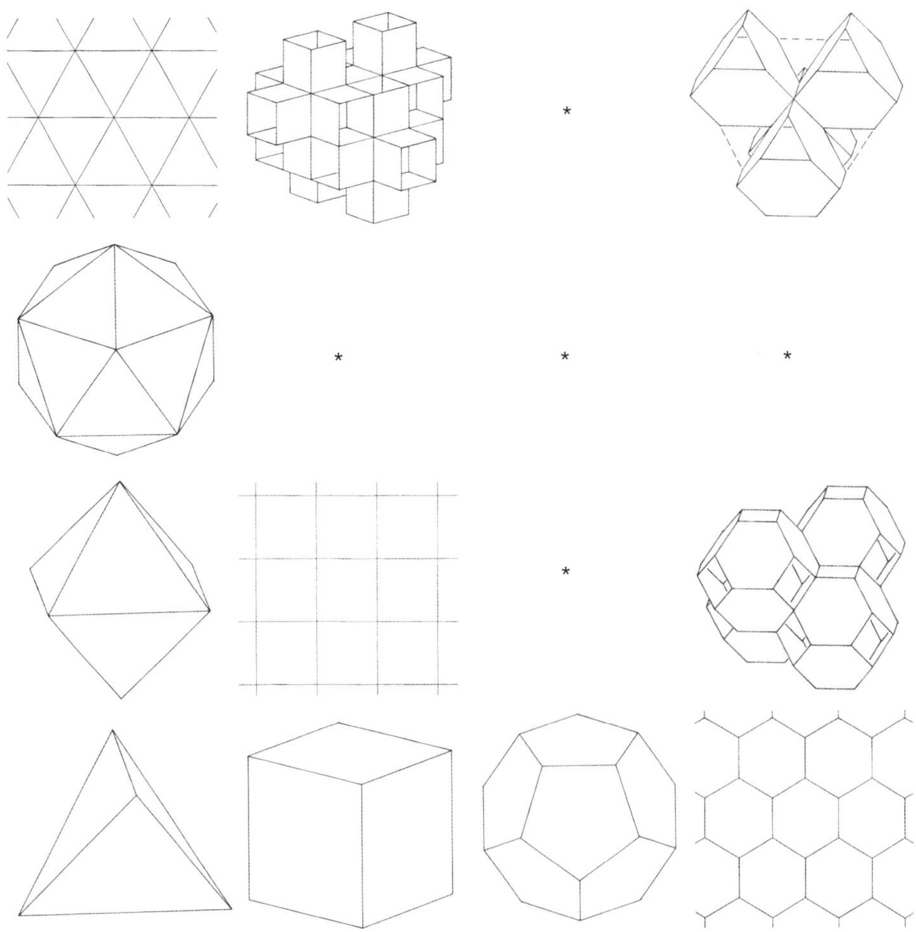

* As symmetry numbers increase vertices become less sharp and polyhedra more spherical until we reach the tilings where the vertices are actually flat. Beyond here vertices can fold up, which requires even symmetry numbers, and we have the skew polyhedra, but that is another subject and perhaps best left for another book !

PLATONIC SOLIDS

Having established our vocabulary of regular and star polygons using an alternative notation in the last chapter, we now need to go up a level of organisation and look at how these polygons can be assembled to produce regular forms. The best known and most regular examples are three dimensional and go back at least to the ancient Greeks.

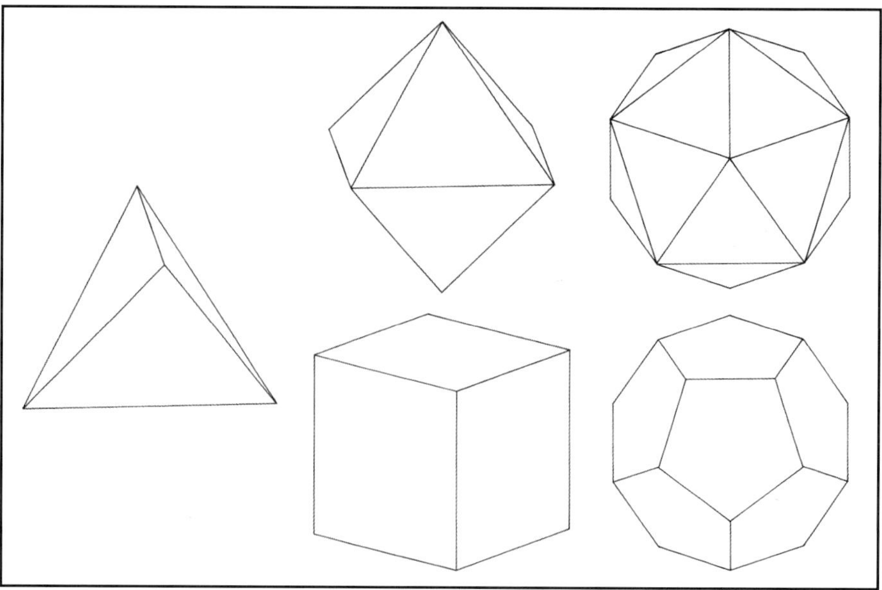

Five in number, the Platonic solids are the basic regular three dimensional geometric forms that can be constructed using regular polygons, supposedly first described by Plato, and yet known to Neolithic man in Scotland. Taking the centres of the faces of any of these as vertex points we can derive another, its dual. They thus comprise two dual pairs, where vertices and faces are interchangeable, and one that is its own dual:

Tetrahedron	$\{3,3\}$	α_3	4 triangle faces, 4 vertices, 6 edges
Cube	$\{4,3\}$	γ_3	6 square faces, 8 vertices, 12 edges
Octahedron	$\{3,4\}$	β_3	8 triangle faces, 6 vertices, 12 edges
Dodecahedron	$\{5,3\}$		12 pentagon faces, 20 vertices, 30 edges
Icosahedron	$\{3,5\}$		20 triangle faces, 12 vertices, 30 edges

The general regular polyhedron is given the Schläfli symbol $\{x,y\}$, indicating that at each vertex there are y x-gons present. The faces thus have symmetry x, the vertices symmetry y and this polyhedron will have a dual given as $\{y,x\}$.

It can be seen that as the symmetry numbers of the Platonic solids increase, the polyhedra become more approximately spherical. Introducing yet more or higher valued polygon faces at each vertex leads to the vertex becoming flat, so that the surface produced by the polygons is not topologically spherical but extends to infinity within a two dimensional plane.

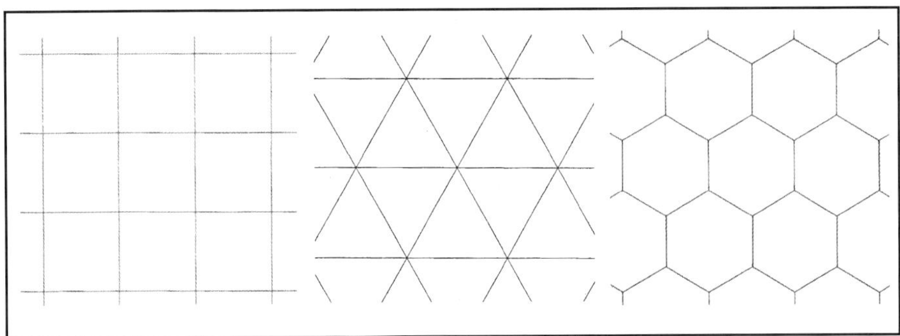

Three in number, the Platonic tilings are the basic regular two dimensional geometric forms that can be constructed using regular polygons. Again taking the centres of the faces of any of these as vertex points we can derive another, its dual. They thus comprise one dual pair, where vertices and faces are interchangeable, and one that is its own dual:

Square tiling $\{4,4\}$ δ_3 4 square faces per vertex

Hexagon tiling $\{6,3\}$ 3 hexagon faces per vertex
Triangle tiling $\{3,6\}$ 6 triangle faces per vertex

The general regular tiling is thus also given the Schläfli symbol $\{x,y\}$, indicating that at each vertex there are y x-gons present. The faces thus have symmetry x, the vertices symmetry y and this tiling will have a dual given as $\{y,x\}$. As with the Platonic polyhedra, the faces are all regular polygons and every vertex is identical.

We thus have a total of five symmetry families comprising the polyhedra and tilings, three with dual pairs, two of them self dual.

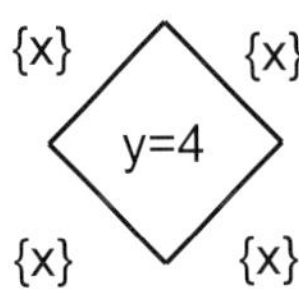

Each vertex of a Platonic form can be represented by a vertex figure that is polygonal in itself, showing the symmetry present there. In general $\{x,y\}$ will have a vertex figure that is a y-gon, each edge of which represents an $\{x\}$.

ARCHIMEDEAN FORMS

Not necessarily first described by Archimedes, these polyhedra and tilings are less pure than the Platonic forms, semi-regular, each having more than one type of face at any vertex, but every vertex still identical. They generally result from the truncation of the Platonic forms and thus involve the truncated polygons we examined in the previous chapter.

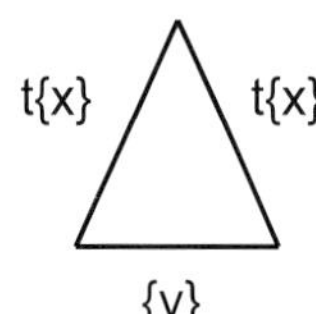

The first truncation of a Platonic form, t{x,y}, will have two faces that are the first truncations of the original polygons t{x}, plus one face that reflects the symmetry of the original vertex {y}. Its vertex figure will thus be an isosceles triangle.

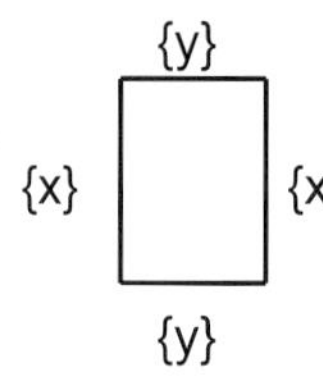

The second truncation t^2{x,y} will be identical to the second truncation of the dual t^2{y,x} and have two of each of the original Platonic forms' polygons at each vertex, hence it is termed quasi-regular. Its vertex figure will therefore be a rectangle, with two {x} polygons and two {y} polygons arranged alternately around it.

These three truncations plus the two original Platonic forms make a first truncation series of forms for each symmetry family. The second truncation at the centre of this series can also be truncated in its turn to create a first and second truncation, starting a second series that eventually results in a less regular form with rhomb faces, except when x=y and the rhombs become squares.

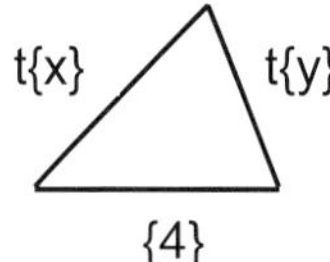

In general tt^2{x,y} will have one t{x}, one t{y} and one square at each vertex, represented by an irregular triangular vertex figure. These forms are known as even faced, since all the faces are truncations with even numbers of edges.

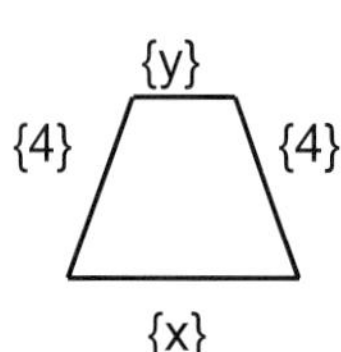

In its turn t^{2}t^2{x,y} will have one {x}, one {y} and two squares at each vertex, the vertex figure forming a trapezium. These are the rhombic forms, from which further truncation leads on to a form with purely rhomb faces, dual to the quasi-regular form.

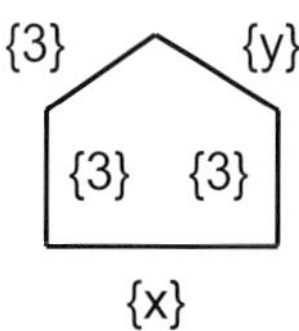

One further symmetrical form can be found for each symmetry family, the snub. This is denoted s{x,y} and is effectively a version of the rhombic form that has been twisted so that each square becomes two regular triangles. Its vertex figure is generally pentagonal with one {x} and one {y}, with three triangles set between them.

One way of representing nearly all the polyhedra or tilings in a symmetry family (except the snubs) is by using Möbius triangles. These are effectively the building blocks of the symmetries present and formed by the lines of symmetry that radiate from each vertex, each face and indeed each edge of a polyhedron or tiling.

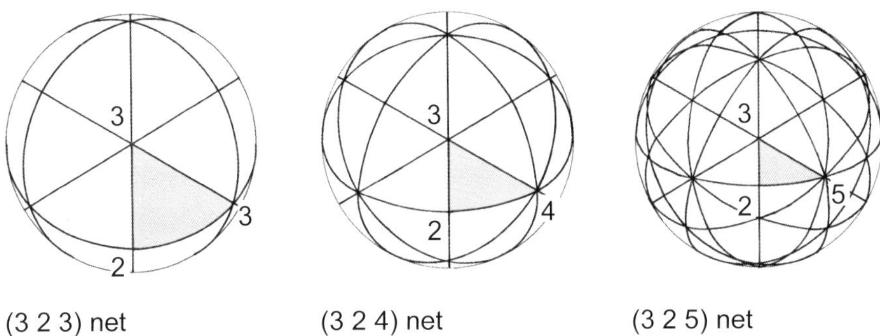

(3 2 3) net   (3 2 4) net   (3 2 5) net

In the polyhedra these lines form a three dimensional spherical net of triangles, whilst in the tilings the net is like them, infinite in extent in two dimensions. The vertices of these nets are numbered with the symmetries that occur there, so that we can find the positions of vertices of any regular or semi-regular form by defining the position within a single Möbius triangle. This is then reflected around the entire spherical or flat net of triangles to define the positions of all the other vertices possible.

In general the vertex of a Platonic form $\{x,y\}$ will occur at every (y) on the net, that of its dual $\{y,x\}$ at every (x), and that of the second truncation of either of these $t^2\{x,y\} = t^2\{y,x\}$ at every (2). These three positions are then given as y| 2 x, x| 2 y and 2| x y respectively in the Wythoff notation, giving us another way to notate the various members within a symmetry family.

Positions along the edges of the Möbius triangles are then notated as vertex positions according to the two vertices at their ends, so that y 2 |x is the first truncation t$\{x,y\}$, x 2 |y is the first truncation t$\{y,x\}$ and x y |2 is the rhombic form $t^2t^2\{x,y\}$.

Two further notations are possible within this Wythoff system: it uses x 2 y| to represent the position at the incentre of the Mobius triangle where the vertex for the even faced form tt$^2\{x,y\}$ can be found. This leaves the notation |x 2 y available to denote the snub s$\{x,y\}$.

Each (x 2 y) Möbius triangle has three colunar triangles (x 2 y'), (x' 2 y) and (x' 2 y') within the same net, where quasitruncated versions of these forms have their vertices denoted using the same notation.

SCHWARZ TRIANGLES

Within our symmetrical nets of small Möbius triangles, we can find larger Schwarz triangles comprising two or more Möbius triangles. These similarly reflect around the nets to cover the sphere or plane with a density equal to the number of Möbius triangles in each Schwarz triangle. Effectively each Möbius triangle is part of two or more different Schwarz triangles.

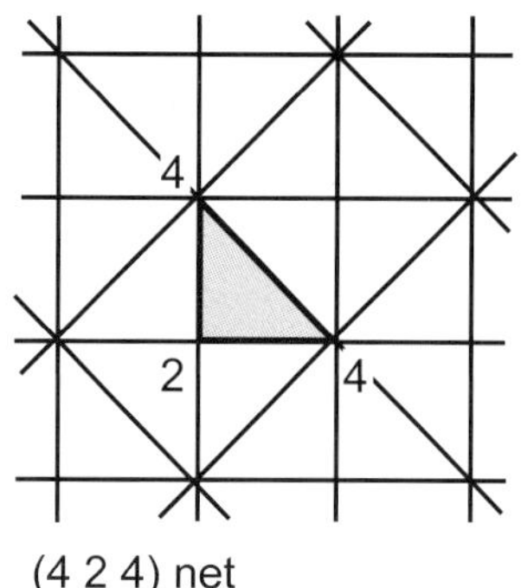

(4 2 4) net

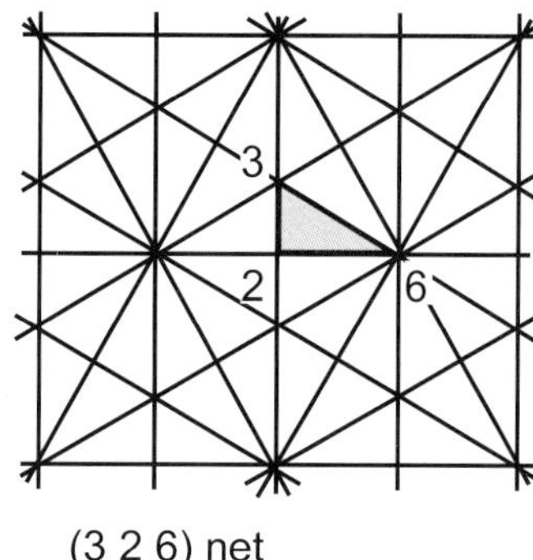

(3 2 6) net

DOUBLE FORMS

The simplest Schwarz triangle is a double one forming an isosceles triangle with two (2) vertices abutting midway along its lower edge. When formed of two (x 2 y) Möbius triangles it will be of the general form (x $^y/_2$ x).

Those polyhedra or tilings with vertices on the axis of the new Schwarz triangle will simply be double versions of certain (x 2 y) forms.
In particular we find that:
$^y/_2$| x x is a double y| 2 x parent form {x,y}
x x |$^y/_2$ is a double 2| x y quasi-regular form t²{x,y}
x $^y/_2$ x| is a double y 2 |x truncated form t{x,y} and
x' $^y/_2$ x'| is a double y 2 |x' quasitruncated form t'{x,y}.

The other positions around the new Schwarz triangle however yield more interesting forms such as the quasi-regular x| $^y/_2$ x, the even faced x $^y/_2$' x'|, three rhombics, x $^y/_2$ |x, x x' |$^y/_2$' and x' $^y/_2$ |x', and three snubs derived from them |x $^y/_2$ x, |x $^y/_2$' x' and |x' $^y/_2$ x'.

The forms which will prove most interesting to us are those of the general form (x $^y/_2$ x) with x = 3, which leaves $^y/_2$ with the potential prospect of being compound. We will thus be looking in the following pages more closely at polyhedra and tilings derived from the Schwarz triangles (3 $^4/_2$ 3), (3 $^5/_2$ 3) and (3 $^6/_2$ 3).

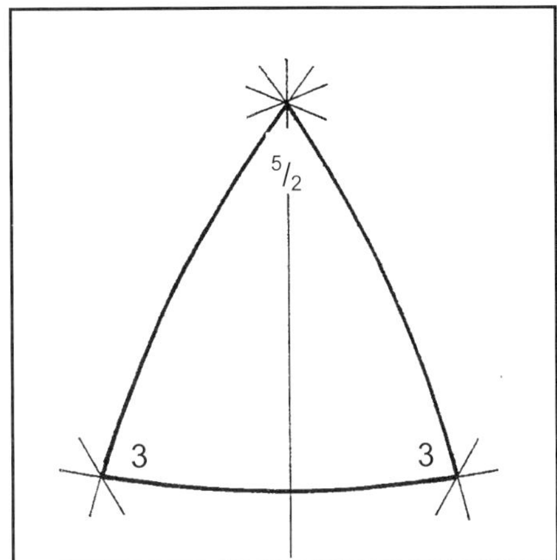
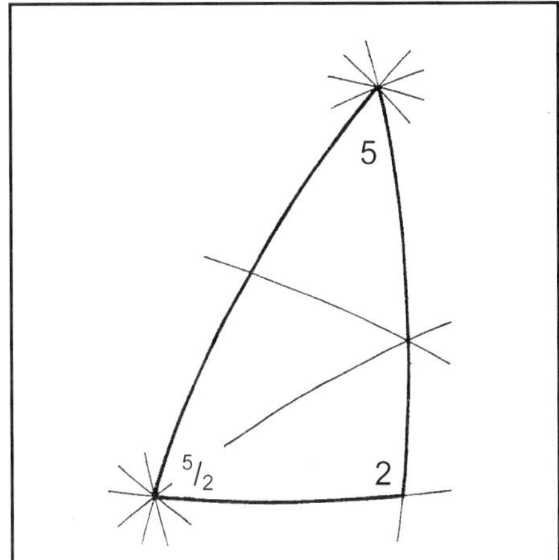

TRIPLE FORMS

When y = 3 a slightly larger Schwarz triangle can be formed by adding a third Möbius triangle to the pair that form (x $^3/_2$ x), generally having the right angled form ($^x/_2$ 2 x). This too reflects around the net, but gives rise to a triple layer of polygons, i.e. with a density of three, each Möbius triangle being in three different Schwarz triangles.

Those in the ($^5/_2$ 2 5) family are generally non-compound star polyhedra, mostly with star polygon {$^5/_2$} faces. They are of particular interest because most are generally accepted as uniform polyhedra and provide the analogues for what will follow in the other cases using compounds derived from ($^6/_2$ 2 6) and ($^4/_2$ 2 4), the star tilings and cross polyhedra respectively.

STAR POLYHEDRA

For the most part already recognised as uniform polyhedra, the star polyhedra of the (3 $^5/_2$ 3) and ($^5/_2$ 2 5) families are now shown purely to illustrate how the star tilings and cross polyhedra that follow in subsequent chapters are their analogues.

It should be noted at this point that other star polyhedra, most notably those in the seven-dense ($^5/_2$ 2 3) family and its derivatives, are not shown here as they do not have interesting analogues in the same way as the (3 $^5/_2$ 3) and ($^5/_2$ 2 5) families. ($^4/_2$ 2 3) and ($^6/_2$ 2 3) are not possible Schwarz triangles!

THE (3 $^5/_2$ 3) FAMILY

3| $^5/_2$ 3

Coxeter 39
Wenninger 70

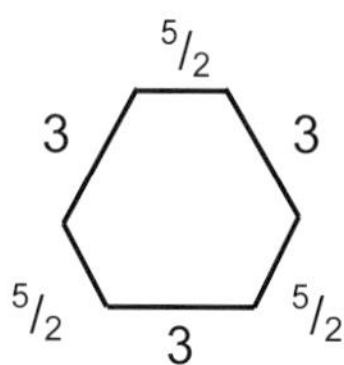

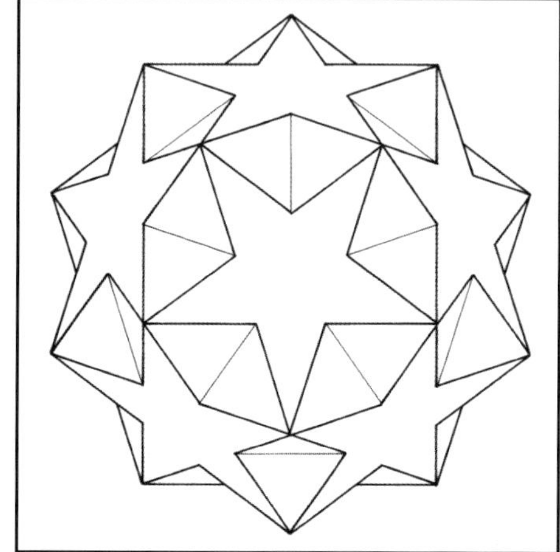

Small Ditrigonal Icosidodecahedron

3 $^5/_2$ |3

Coxeter 40
Wenninger 71

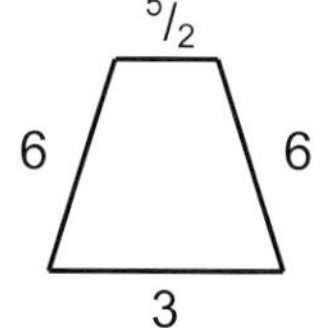

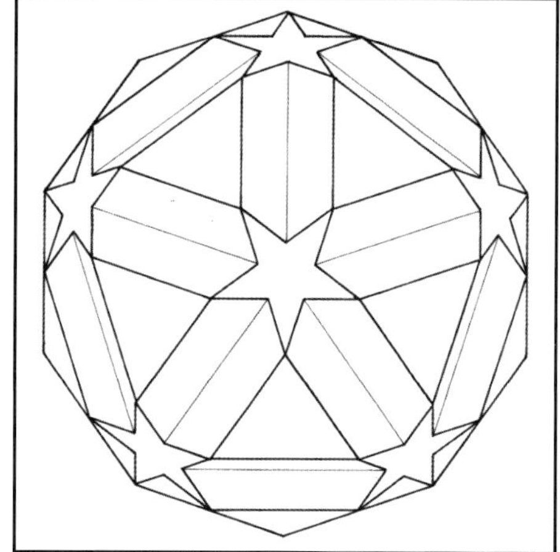

Small Icosicosidodecahedron

|3 $^5/_2$ 3

Coxeter 41
Wenninger 110

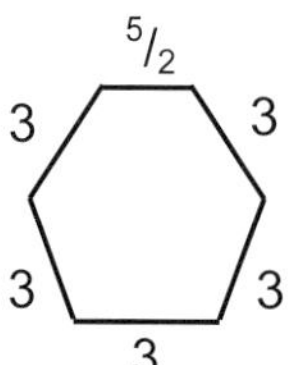

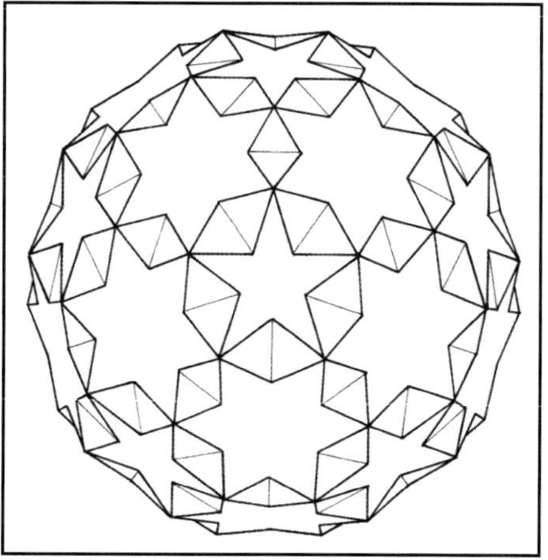

Small Snub Icosicosidodecahedron

3 $^5/_3$ $^3/_2$|

Coxeter -
Wenninger -

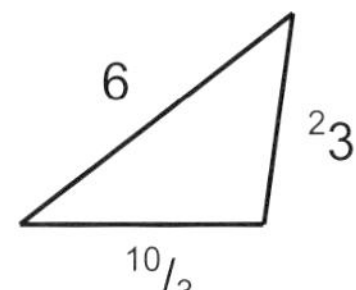

Quasiquasitruncated Small Ditrigonal Icosidodecahedron

Ditrigonal vertex figure type 2 (p.84)

Fits within shell of 3 5 |$^5/_3$

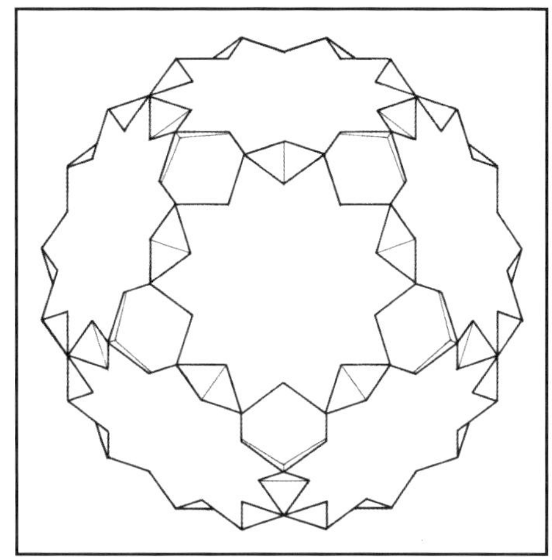

3 $^3/_2$ |$^5/_3$

Coxeter 85
Wenninger 106

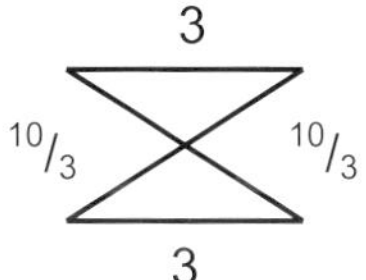

Great Icosihemidodecahedron

Fits within shell of 2| 3 $^5/_2$

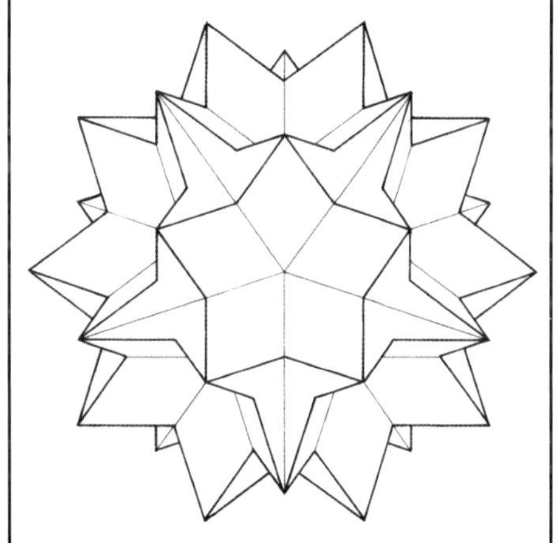

|3 $^5/_3$ $^3/_2$

Coxeter Table 6
Wenninger -

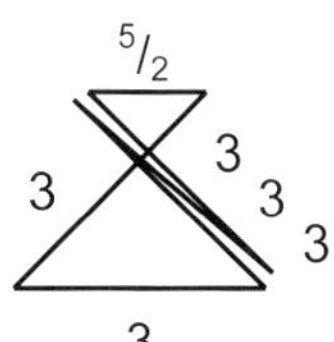

Quasisnub Icosicosidodecahedron

Fits within shell of 5| 2 $^5/_2$

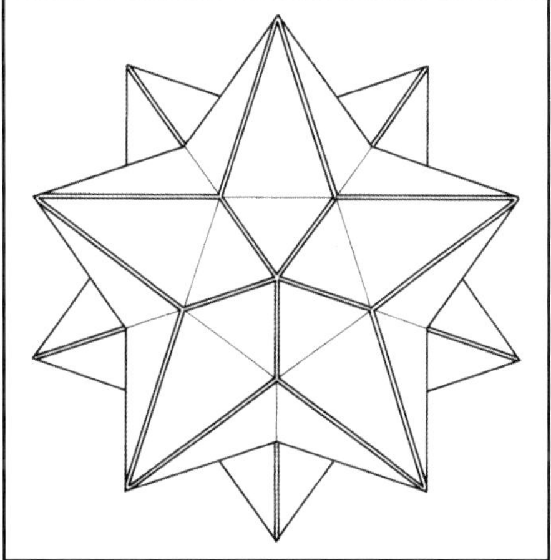

THE (3 $^5/_2$ 3) FAMILY

It will be noticed that when there is not a '2' in the Schwarz triangle for a particular family, the vertex figures are slightly more complex. The two shown in this section as from Coxeter's Table 6 of 'degenerate cases of Wythoff's construction' have vertex figures that fit within the larger vertex figure of the Small Stellated Dodecahedron 5| 2 $^5/_2$.

The snubs typically have six faces per vertex - three triangles interlaced with one of each of the defining polygons for the family (when there is a '2' this is simply a digon forming the edge between the paired triangles).

The two quasirhombihedra shown below and overleaf have typically twisted vertex figures resulting from the inclusion of an inverse polygon in the first part of the symbol. Similarly the quasisnub polyhedra below them have two such twists resulting from two inverse polygons being present.

$3 \, ^5/_3 \, |^3/_2 = \, ^3/_2 \, ^5/_2 \, |^3/_2$

Coxeter Table 6
Wenninger -

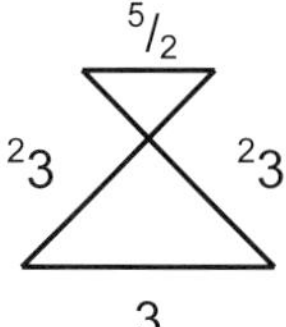

Small Quasicosicosidodecahedron

Fits within shell of 5| 2 $^5/_2$

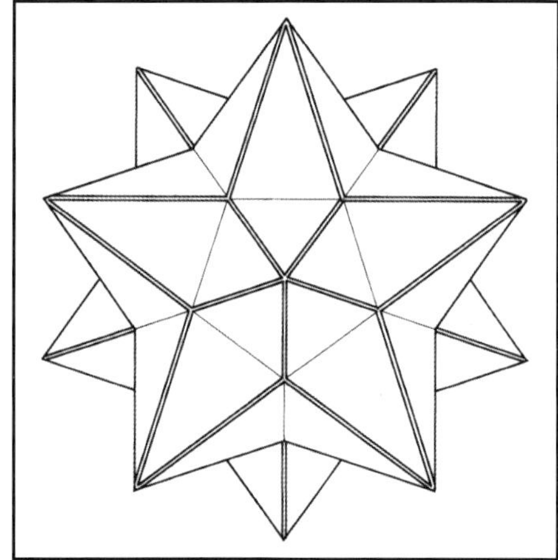

$|^3/_2 \, ^5/_2 \, ^3/_2$

Coxeter 91
Wenninger 118

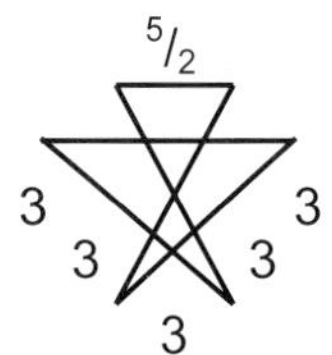

Small Inverted Retrosnub
Icosicosidodecahedron

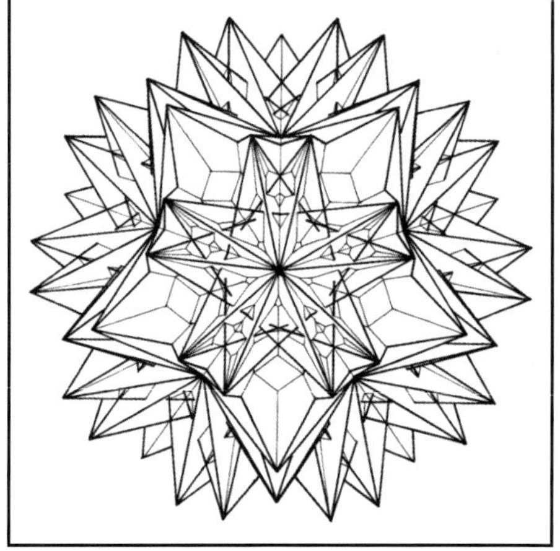

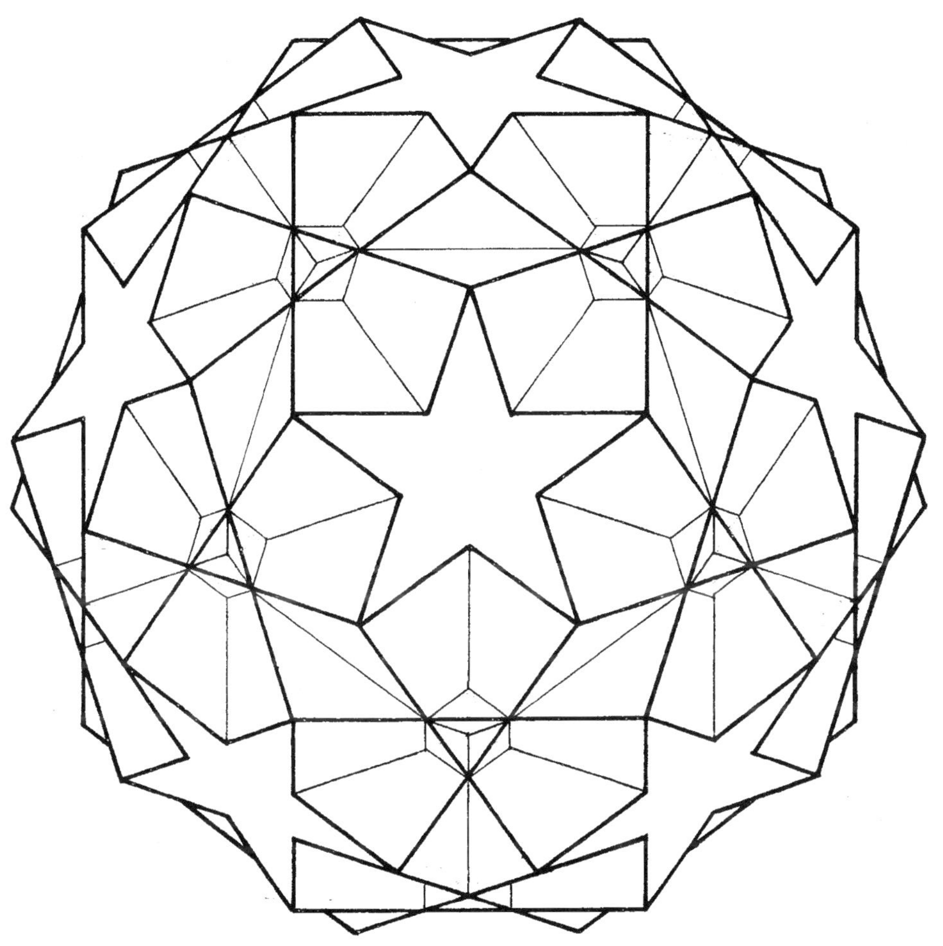

$^5/_2$ 5 |2 $t^2t^2\{5,{}^5/_2\}$

THE $(^5/_2\,2\,5)$ FAMILY

$^5/_2|\,2\,5$ $\{5,^5/_2\}$

Coxeter 44
Wenninger 21

Great Dodecahedron

Inscription of Icosahedron $\{3,5\}$

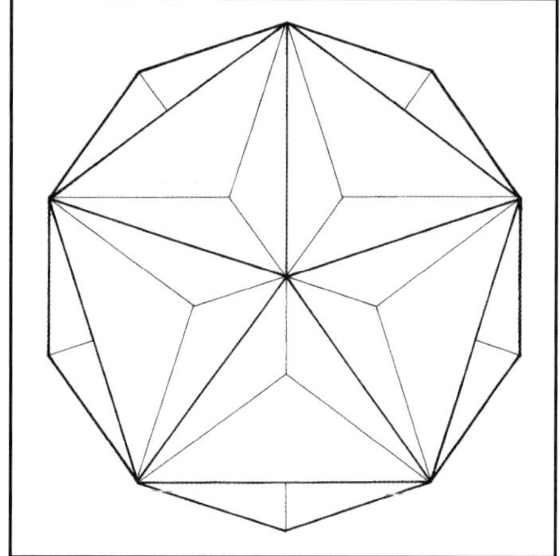

$^5/_2\,2\,|5$ $t\{5,^5/_2\}$

Coxeter 47
Wenninger 75

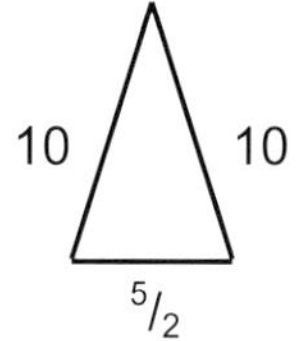

Truncated Great Dodecahedron

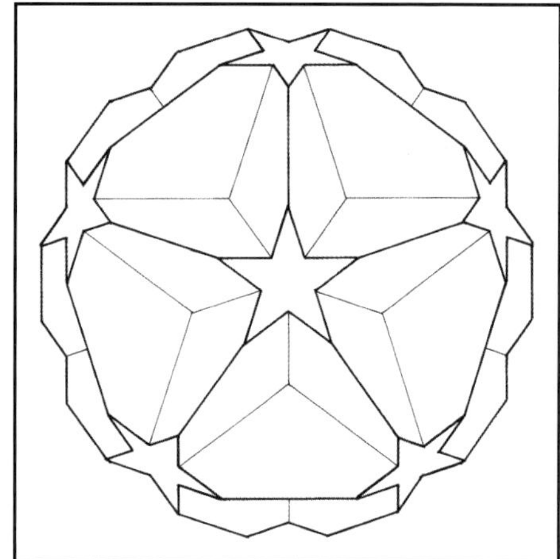

$^5/_2\,2\,|^5/_4$ $t\{5',^5/_2\}$

Coxeter Table 6
Wenninger -

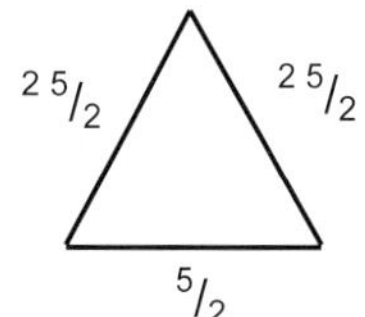

Quasitruncated Great Dodecahedron

Three coincident Great Stellated
Dodecahedra $\{^5/_2,3\}$

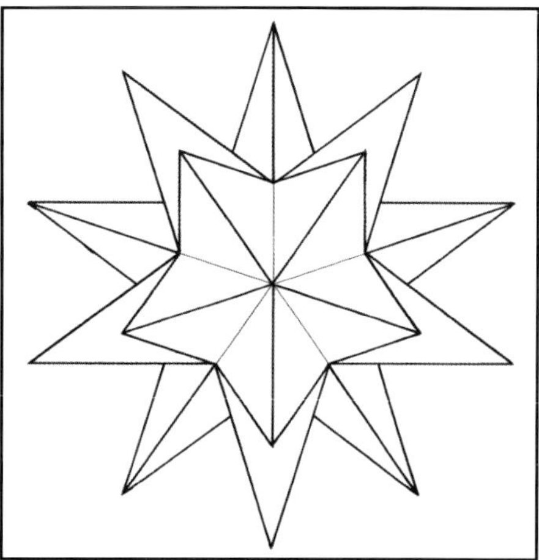

$5|\ 2\ ^5/_2 \qquad \{^5/_2,5\}$

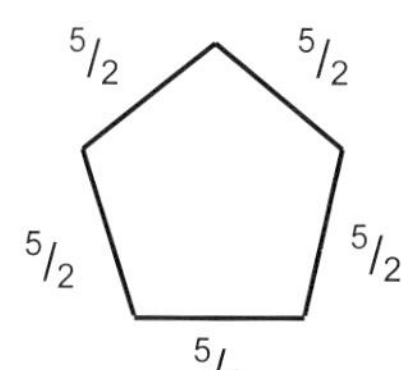

Coxeter 43
Wenninger 20

Small Stellated Dodecahedron

Stellation of Dodecahedron {5,3}

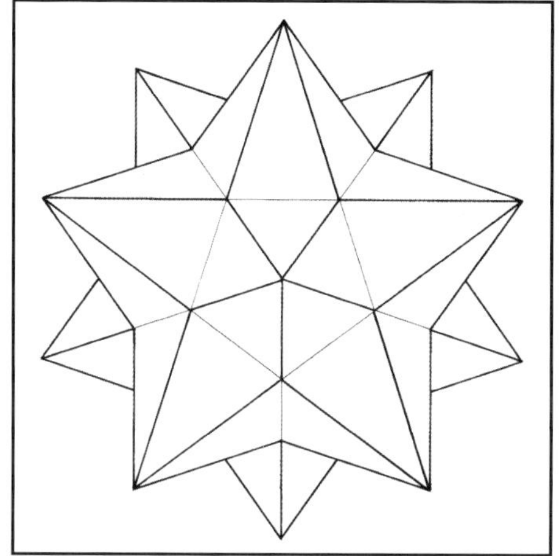

$5\ 2\ |^5/_2 \qquad t\{^5/_2,5\}$

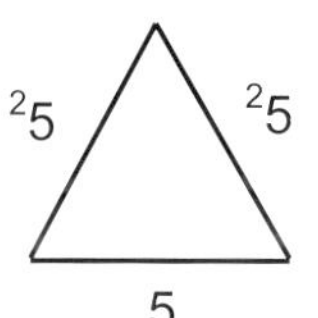

Coxeter Table 6
Wenninger -

Truncated Small Stellated
Dodecahedron

Three coincident Dodecahedra {5,3}

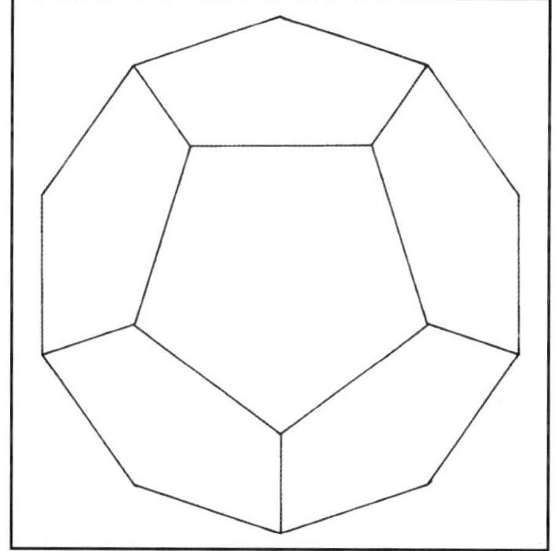

$5\ 2\ |^5/_3 \qquad t\{^5/_2{}',5\}$

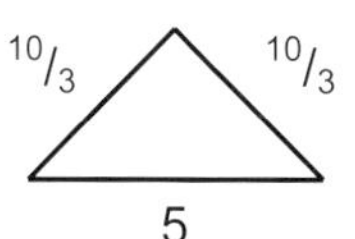

Coxeter 74
Wenninger 97

Quasitruncated Small Stellated
Dodecahedron

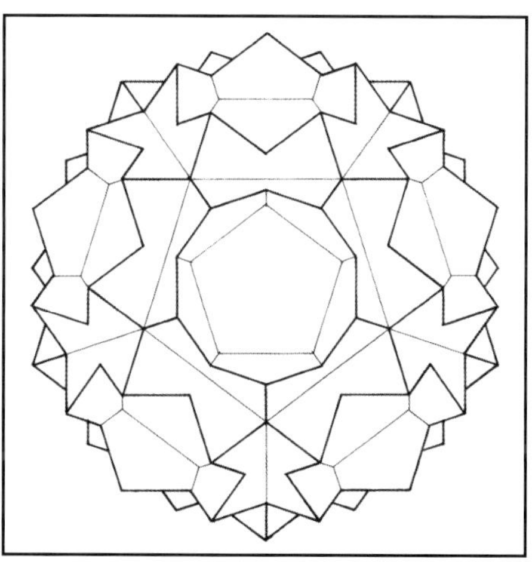

THE ($^5/_2$ 2 5) FAMILY

2| $^5/_2$ 5 $t^2\{5,^5/_2\}$

Coxeter 45
Wenninger 73

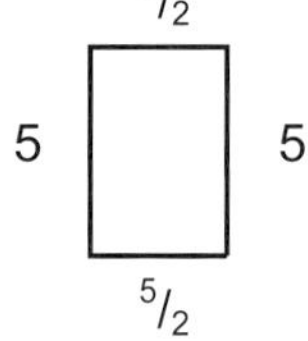

Dodecadodecahedron

Second truncation of either $\{5,^5/_2\}$ or $\{^5/_2,5\}$, quasi-regular form at centre of first truncation series

Has four possible (quasi) truncations as below and opposite

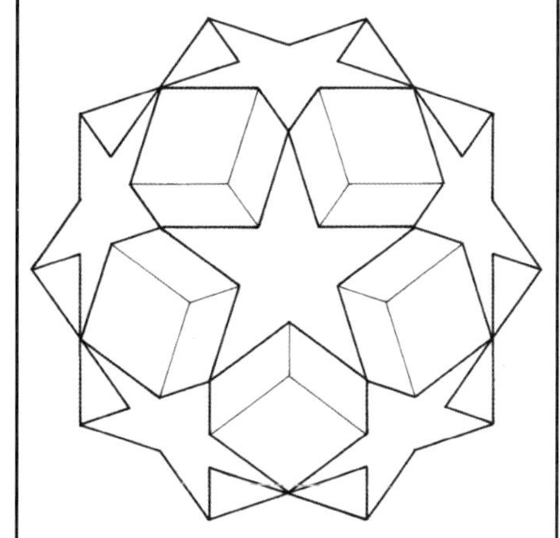

5 2 $^5/_2$| $tt^2\{5,^5/_2\}$

Coxeter -
Wenninger -

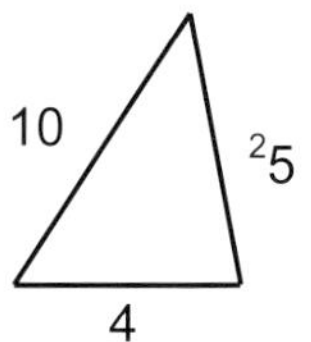

Truncated Dodecadodecahedron

Ditrigonal vertex figure type 2 (p.82)

Fits within shell of 3 5 |2

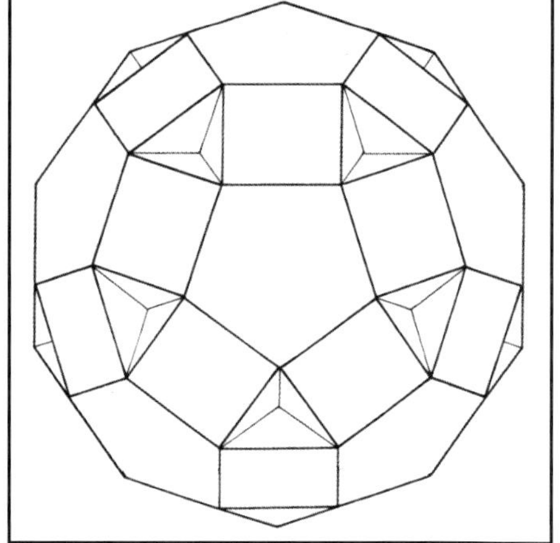

$^5/_4$ 2 $^5/_2$| $tt^2\{5',^5/_2\}$

Coxeter Table 6
Wenninger -

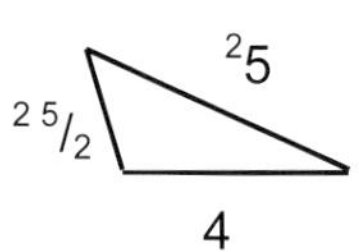

Pentaquasitruncated Dodecadodecahedron

Ditrigonal vertex figure type 3 (p.82)

Fits within shell of 5 $^5/_3$ |2

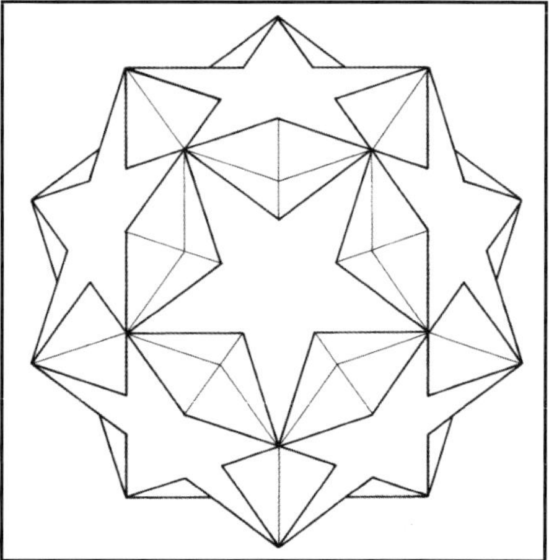

DITRIGONAL VERTEX FIGURES

The even faced polyhedra x y z| that result when x, y and z are all different and one of them has an even denominator give rise to ditrigonal vertex figures, the subject of a later chapter. In essence a double polygon is present twice at each vertex and generates a vertex figure of type 2 with six faces present, comprising two interlinked triangles. We have already seen 3 $^5/_3$ $^3/_2$| on p.23 and two more present themselves here, 5 2 $^5/_2$| opposite and $^5/_4$ 2 $^5/_3$| below.

When there are two even denominators present we get two double faces in a vertex figure of type 3, that fit within the shell of the Small Ditrigonal Icosidodecahedron 3| $^5/_2$ 3. One appears opposite below $^5/_4$ 2 $^5/_2$| and these are all included in Coxeter's Table 6 of 'degenerate cases of Wythoff's construction'.

5 2 $^5/_3$| tt^2\{5,$^5/_2$'\}

Coxeter 75
Wenninger 98

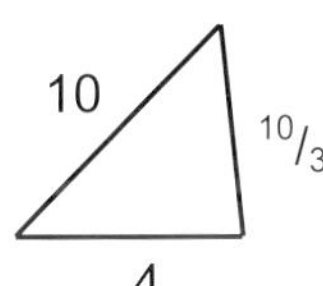

Quasitruncated
Dodecadodecahedron

Only (quasi) truncation of the
Dodecadodecahedron that does not
involve even denominator polygons
and thus double polygon faces

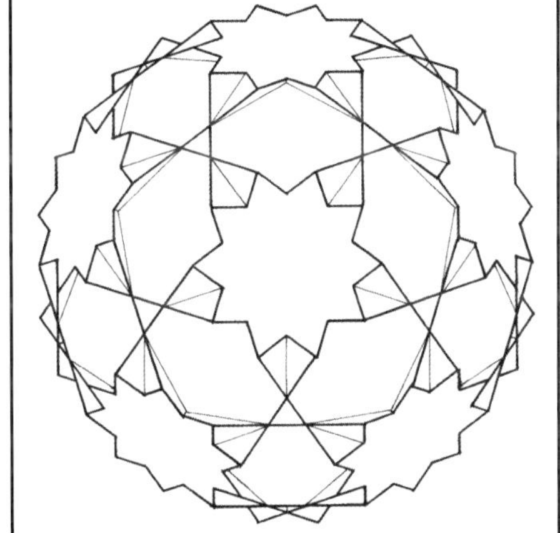

$^5/_4$ 2 $^5/_3$| tt^2\{5',$^5/_2$'\}

Coxeter -
Wenninger -

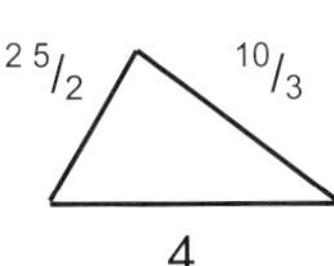

Quasiquasitruncated
Dodecadodecahedron

Ditrigonal vertex figure type 2 (p.82)

Fits within shell of 3 $^5/_2$ |$^5/_3$

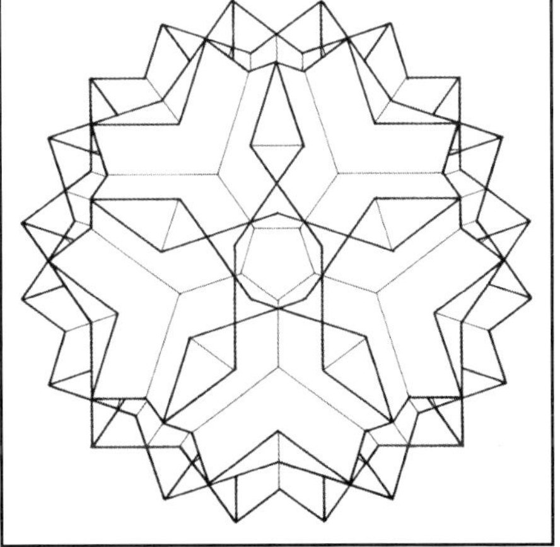

$^5/_2$ 5 |2 $t^2t^2\{5,^5/_2\}$

Coxeter 48
Wenninger 76

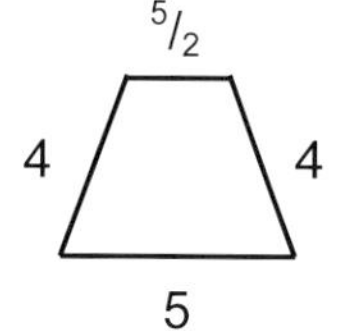

Rhombidodecadodecahedron

Second truncation of quasi-regular
Dodecadodecahedron

Theoretical centre of second
truncation series

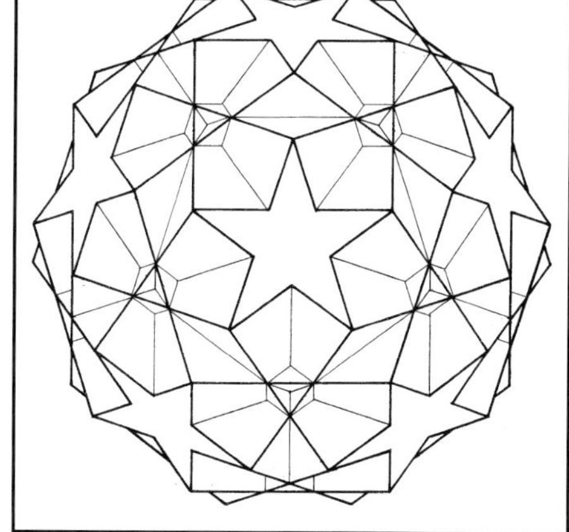

|$^5/_2$ 2 5

Coxeter 49
Wenninger 111

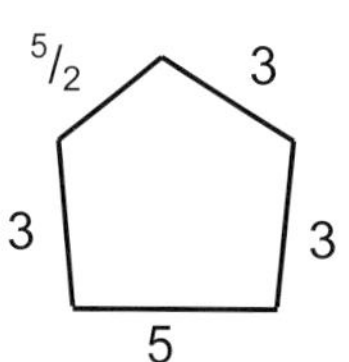

Snub Dodecadodecahedron

Twisted version of above
Rhombidodecadodecahedron

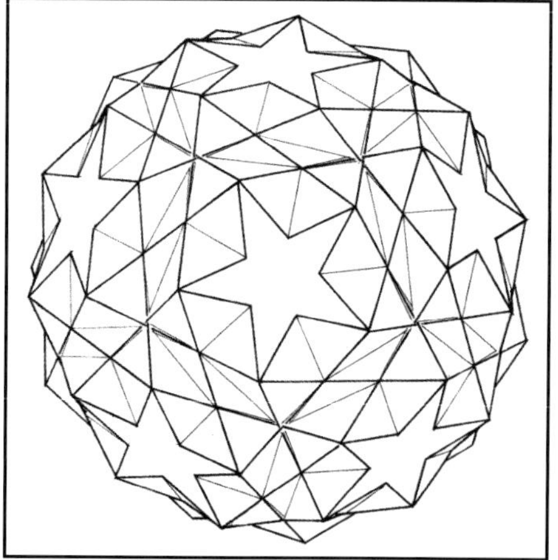

RHOMBIHEDRA & SNUBS

Opposite and below are shown the rhombihedral and quasirhombihedral polyhedra for this family respectively, each with its associated snub form. The latter's vertex figures do not need to show the redundant digon {2} between the two adjacent triangles {3}, as this is simply the edge between those two triangles.

$^5/_3$ 5 |2 t^2t^2{5,$^5/_2$'}

Coxeter Table 6
Wenninger -

Quasirhombidodecadodecahedron

Fits within shell of 3| $^5/_2$ 3

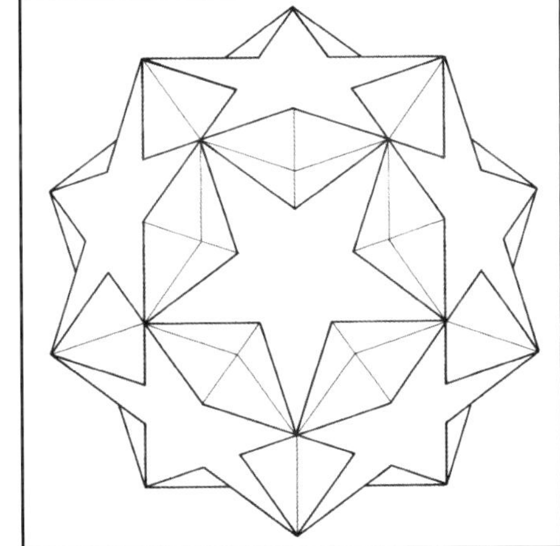

|$^5/_3$ 2 5

Coxeter 76
Wenninger 114

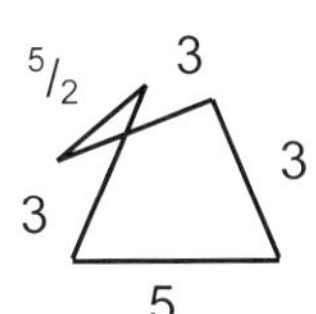

Quasisnub Dodecadodecahedron

Twisted version of above
Quasirhombidodecadodecahedron

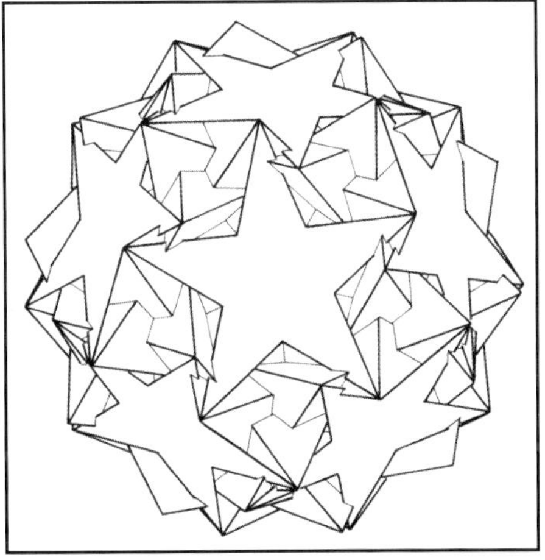

REFERENCES

Coxeter, H.S.M. 1973 *Regular Polytopes* Dover, New York

Coxeter, H.S.M., Longuet-Higgins, M.S. & Miller, J.C.P. 1953 *Uniform Polyhedra* Phil. Trans. R. Soc. Lond. A 246, 401-449

Critchlow, K. 1969 *Order in Space* Thames and Hudson

Cundy, H.M. & Rollett, A.P. 1951 *Mathematical Models* Oxford U P

Grünbaum, B. & Shephard, G.C. 1989 *Tilings and Patterns; An Introduction* W.H.Freeman

Holden, A. 1971 *Shapes, Space, and Symmetry* Columbia U P

Open University 1994 *Tilings* M336 Block 1, Unit IB1

Skilling, J. 1974 *The Complete Set of Uniform Polyhedra* Phil. Trans. R. Soc. Lond. A 278, 111-135

Taylor, P. 1995 *Additions to the Uniform Polyhedra* Nattygrafix

Taylor, P. 1997 *The Complete? Polygon* Nattygrafix

Taylor, P. 1998 *Incomplete Tilings* Nattygrafix

Taylor, P. 1999 *The Simpler? Polyhedra* Nattygrafix

Taylor, P. 2000 *The Star & Cross Polyhedra* Nattygrafix

Wenninger, M.J. 1971 *Polyhedron Models* Cambridge U P

Wenninger, M.J. 1983 *Dual Models* Cambridge U P

We now come to the prospect of exploring new tilings made possible by creating two new Schwarz triangles from within the (3 2 6) net of Möbius triangles. They are the twofold (3 $^6/_2$ 3) and threefold ($^6/_2$ 2 6) Schwarz triangles and the results follow in pages set out to match those previously seen for the twofold (3 $^5/_2$ 3) and threefold ($^5/_2$ 2 5) Schwarz triangles of the star polyhedra, to which they are direct analogues.

They employ the polygon $\{^6/_2\}$ as described by the alternative notation adopted in this book, i.e. this star polygon appears as the 'Star of David' comprising two out of phase triangles. If the currently held view of $\{^6/_2\}$ were to be adopted we would be dealing with a double coincident triangle, the result of truncating a $\{^3/_2\}$ and all the new tilings that might present themselves would simply be doubled up versions of the (3 2 6) tilings, which we already know about, all rather boring.

The interpretation of these polygons in the alternative notation introduces an 'out of phase' element that allows these tilings to fill the plane in new and interesting ways, which are effectively kaleidoscopic. However it will be seen that the results are consistent and even in terms of filling the plane exactly twice or thrice over as applicable.

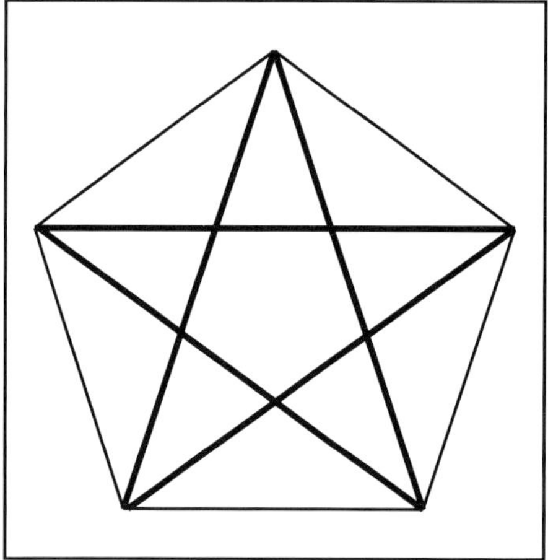

Inscribed Pentagon

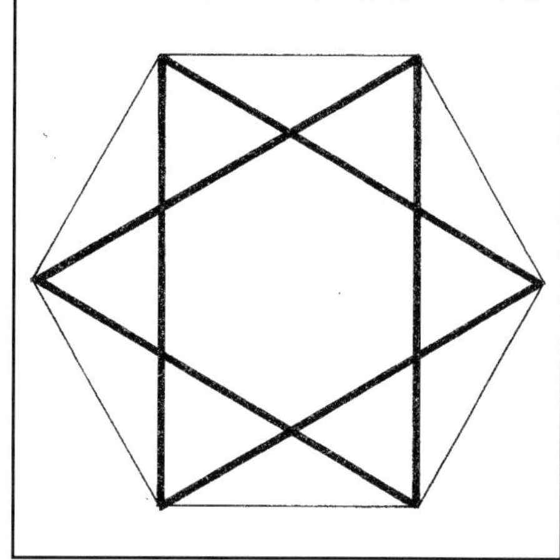

Inscribed Hexagon

The star polygon $\{^6/_2\}$ can be derived from a hexagon $\{6\}$ in the same way that the star polygon $\{^5/_2\}$ is derived from a pentagon $\{5\}$. This can be achieved by either inscribing within or stellating without as shown above and below respectively.

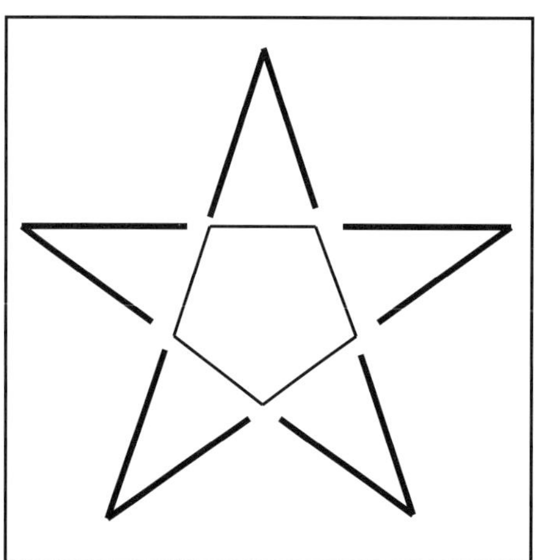

Stellated Pentagon

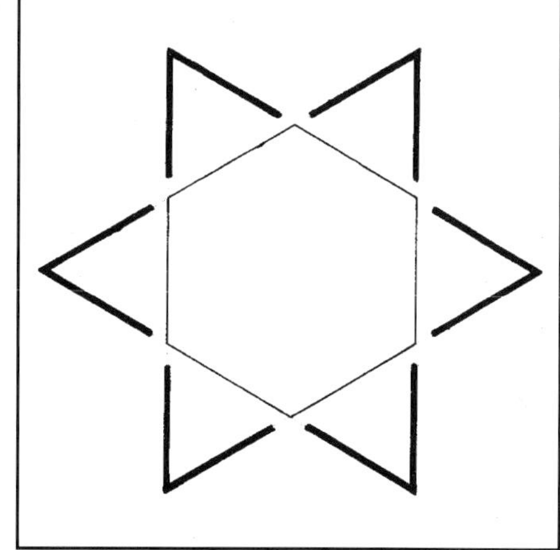

Stellated Hexagon

The new $\{^6/_2\}$ star polygon can be seen as two triangles overlapping each other and is notated as $2\{3\}$ by some geometers. The acceptance of this 'Star of David' face into the polygonal vocabulary leads to two consistent families of tilings, each based on a particular Schwarz triangle that can be found within the lattice of (3 2 6) Möbius triangles.

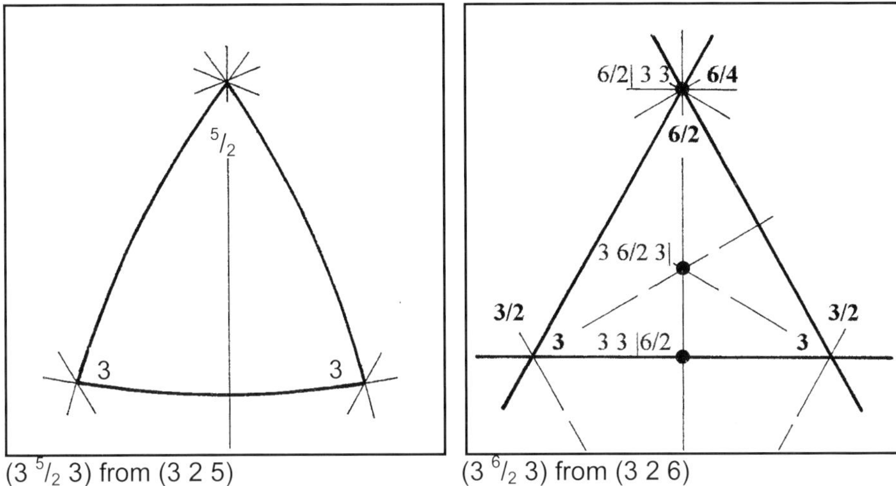

(3 $^5/_2$ 3) from (3 2 5)   (3 $^6/_2$ 3) from (3 2 6)

The (3 $^5/_2$ 3) Schwarz triangle in the polyhedra, derived from two (3 2 5) icosahedral Möbius triangles, can be paralleled by a (3 $^6/_2$ 3) Schwarz triangle derived from the (3 2 6) tiling family with both families having density 2 as each Schwarz triangle contains two Möbius triangles.

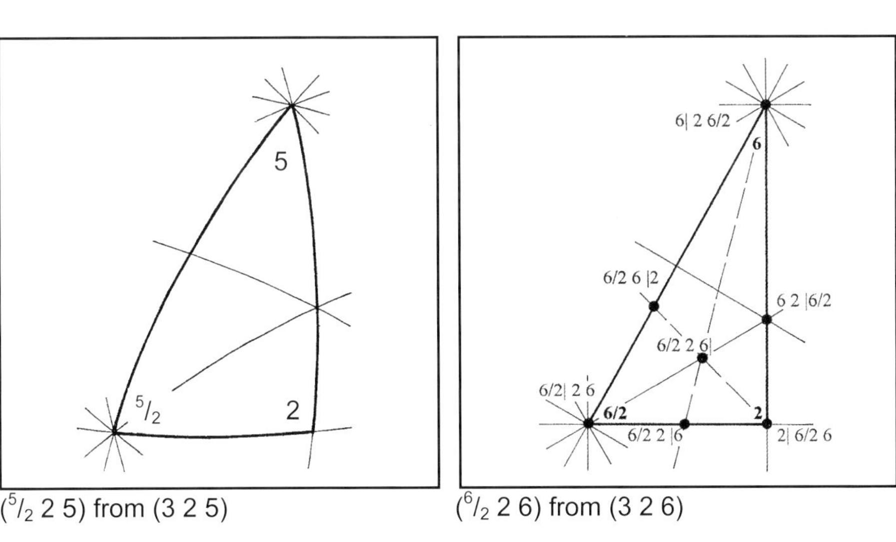

($^5/_2$ 2 5) from (3 2 5)   ($^6/_2$ 2 6) from (3 2 6)

The ($^5/_2$ 2 5) Schwarz triangle can be derived from three (3 2 5) icosahedral Möbius triangles leading to a star polyhedra family with density 3 (three of the latter comprising the former). Similarly a ($^6/_2$ 2 6) Schwarz triangle can be derived from the (3 2 6) tiling Möbius triangle leading to a complete set of parallel inter-related tilings with density 3.

THE (3 $^6/_2$ 3) FAMILY

3| $^6/_2$ 3

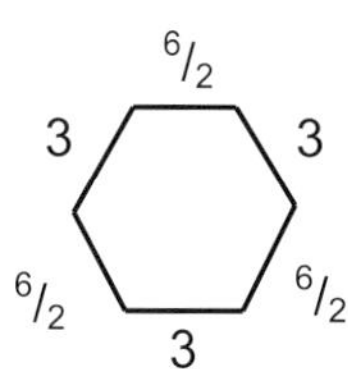

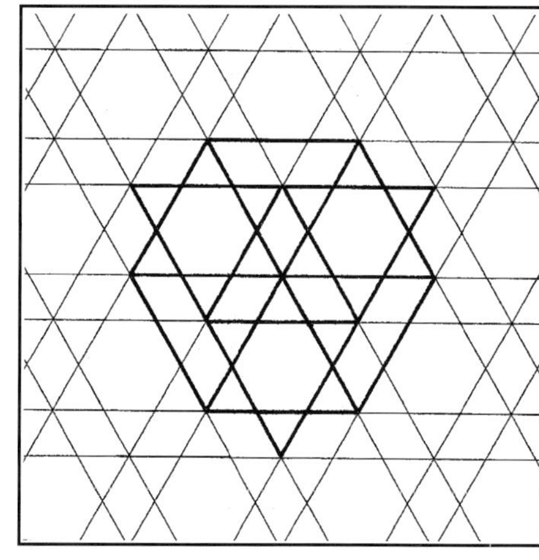

Analogue of Small Ditrigonal
Icosidodecahedron

3 $^6/_2$ |3

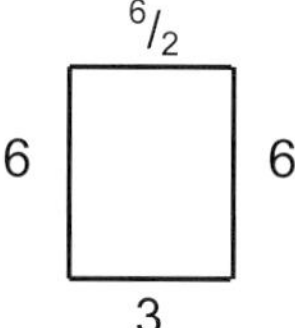

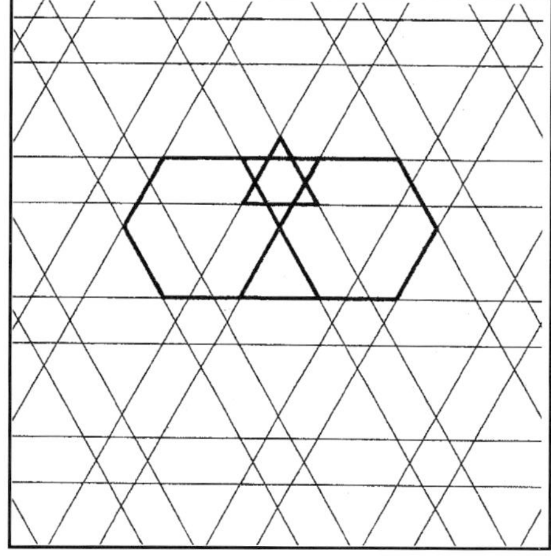

Analogue of Small
Icosicosidodecahedron

Density 2 most apparent: every
triangle concentric with a hexagon

|3 $^6/_2$ 3

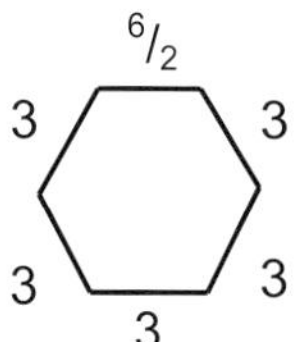

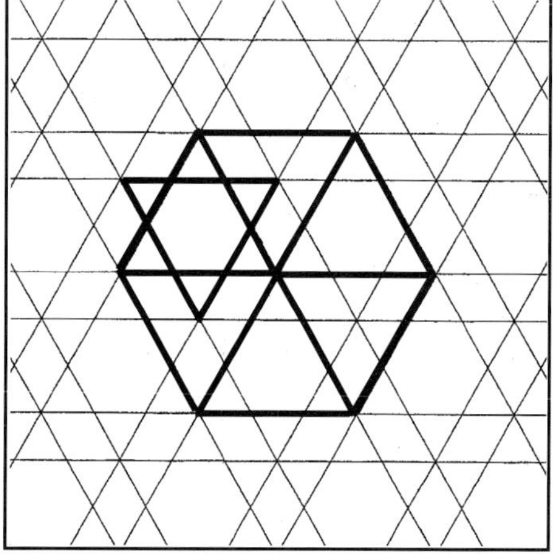

Analogue of Small Snub
Icosicosidodecahedron

Missing from 'Incomplete Tilings'
but same appearance as 3| $^6/_2$ 3

3 $^6/_4$ $^3/_2$|

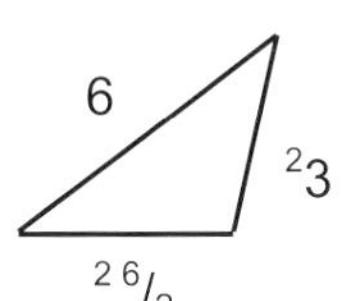

Analogue of Quasiquasitruncated
Small Ditrigonal Icosidodecahedron

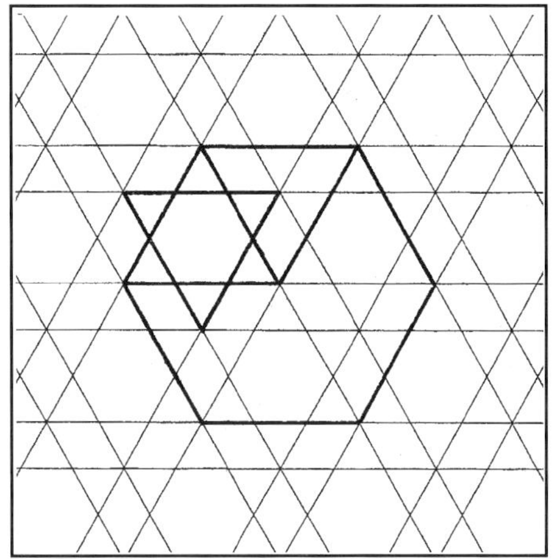

3 $^3/_2$ |$^6/_4$

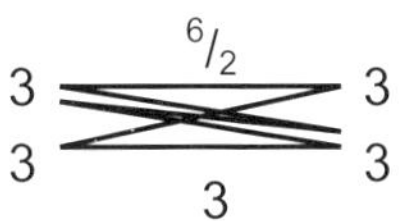

Analogue of Great
Icosahemidodecahedron

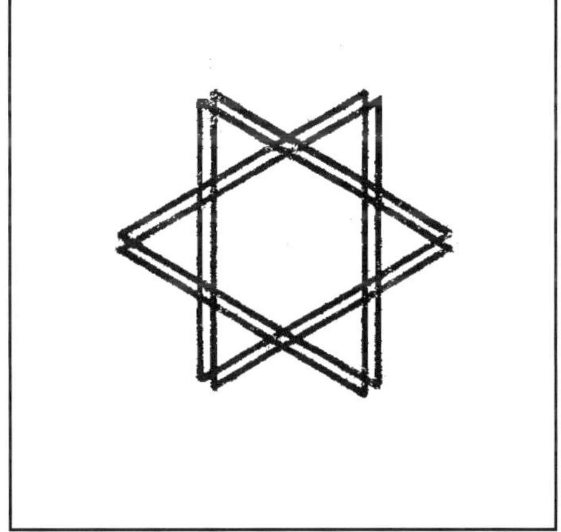

|3 $^6/_4$ $^3/_2$

Analogue of Quasisnub
Icosicosidodecahedron

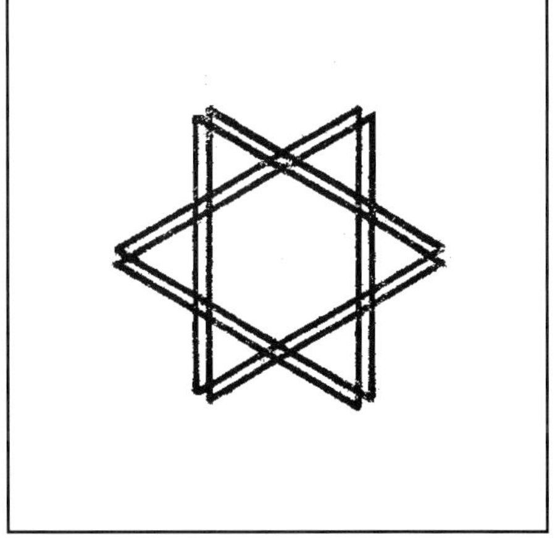

THE (3 $^6/_2$ 3) FAMILY

The last four predicted vertices of this family (two below and two overleaf) all have the property that they do not get to fill the plane. Rather they simply close up on themselves to produce finite assemblies of polygons that have just six vertices, those of a $\{^6/_2\}$ Star of David. The stars included within each vertex figure ensure that there are six rather than just three vertices in each case.

$3\ ^6/_4\ |^3/_2$ or
$^3/_2\ ^6/_2\ |^3/_2$

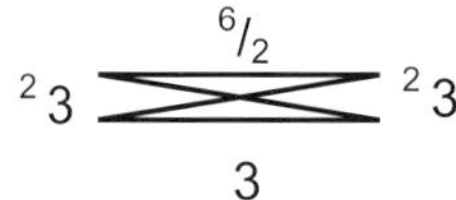

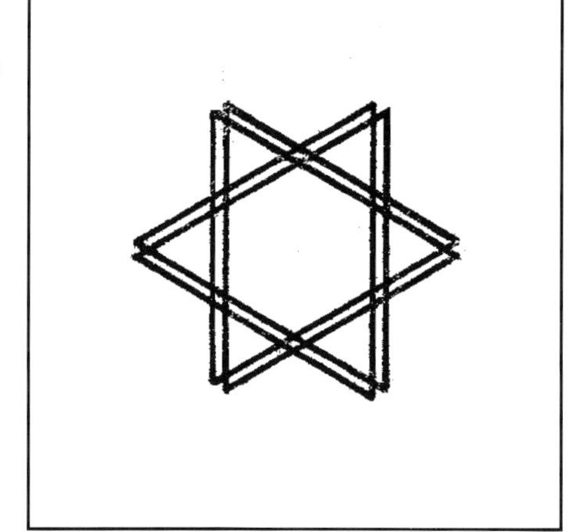

Analogue of Small
Quasicosicosidodecahedron

$|^3/_2\ ^6/_2\ ^3/_2$

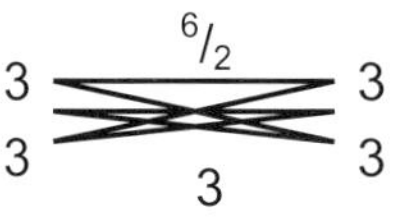

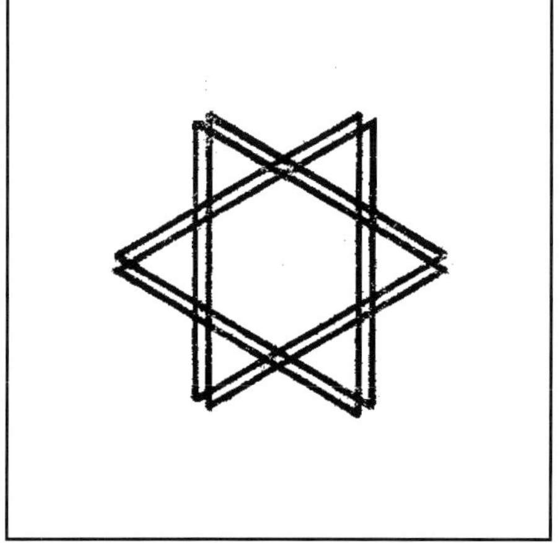

Analogue of Small Inverted
Retrosnub Icosicosidodecahedron

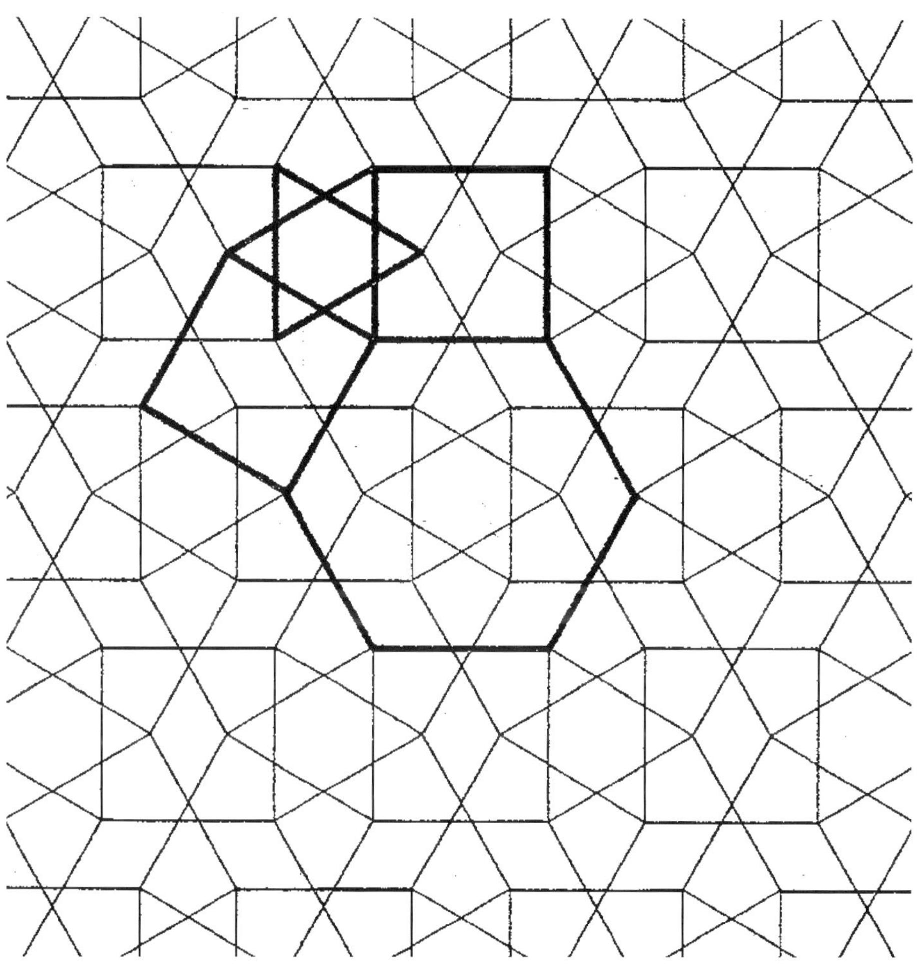

$^6/_2$ 6 |2 t^2t^2{6,$^6/_2$}

THE ($^6/_2$ 2 6) FAMILY

$^6/_2|$ 2 6   {6,$^6/_2$}

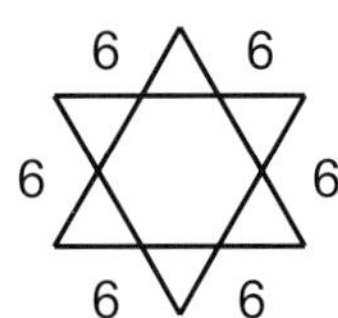

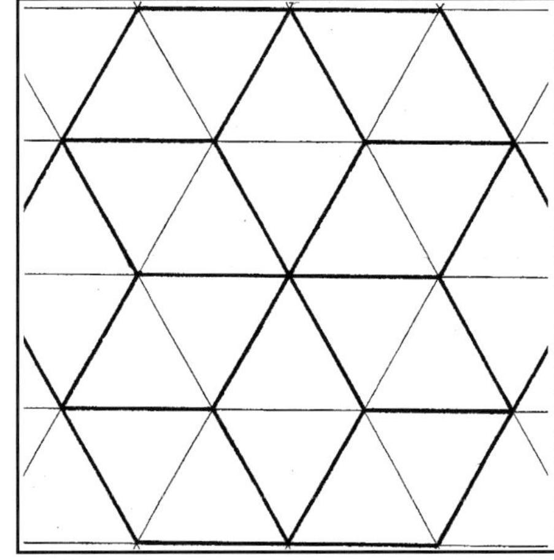

Each apparent triangle here is part of
three overlapping hexagons, six of
which meet at any given vertex

Analogue of Great Dodecahedron

$^6/_2$ 2 |6   t{6,$^6/_2$}

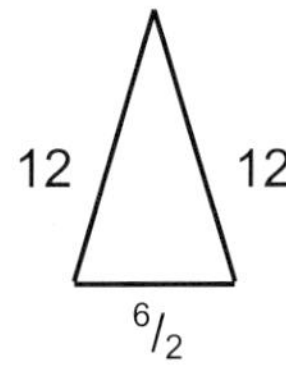

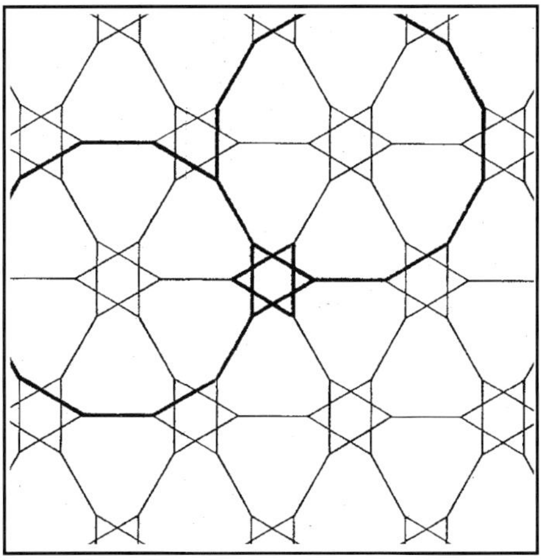

Analogue of Truncated Great
Dodecahedron

$^6/_2$ 2 |$^6/_5$ t{6',$^6/_2$}

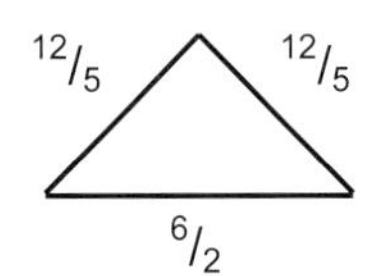

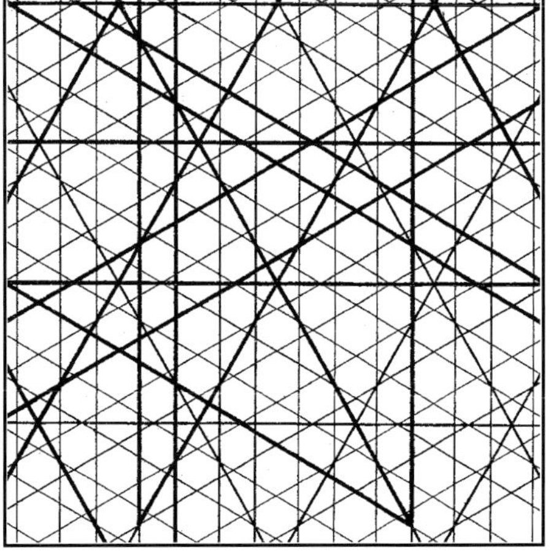

Analogue of Quasitruncated Great
Dodecahedron

6| 2 $^6/_2$ $\{^6/_2,6\}$

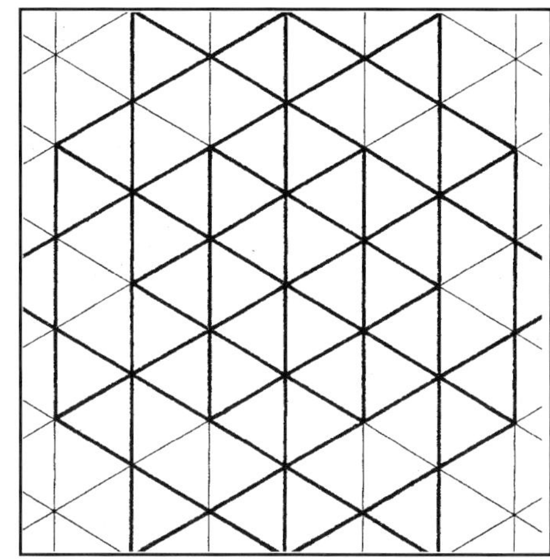

Analogue of Small Stellated
Dodecahedron

6 2 |$^6/_2$ t$\{^6/_2,6\}$

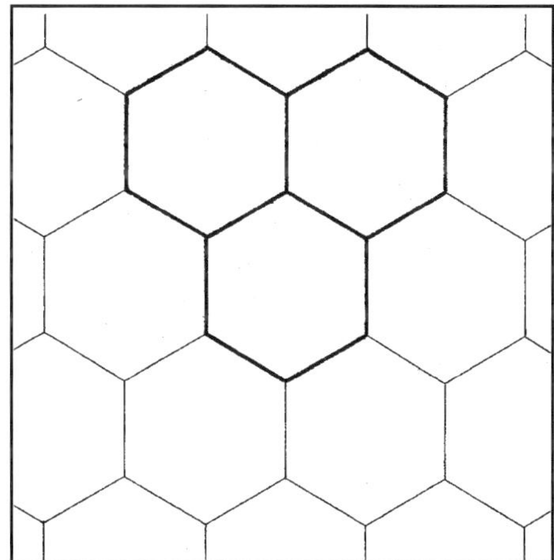

Here we have a triple Hexagon Tiling

Analogue of Truncated Small
Stellated Dodecahedron, which is
itself a triple Dodecahedron

6 2 |$^6/_4$ t$\{^6/_2{'},6\}$

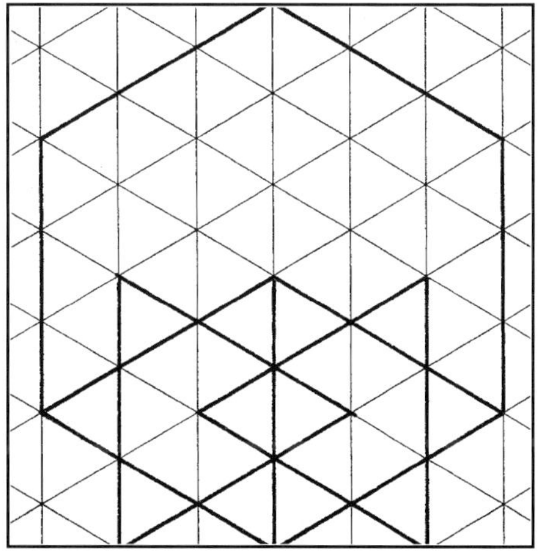

Analogue of Quasitruncated Small
Stellated Dodecahedron

THE $(^6/_2\ 2\ 6)$ FAMILY

$2|\ ^6/_2\ 6 \qquad t^2\{6,^6/_2\}$

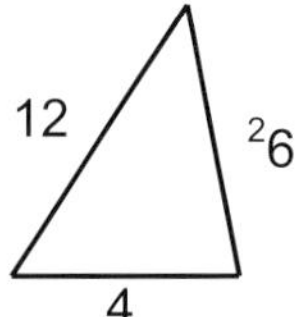

Analogue of Dodecadodecahedron

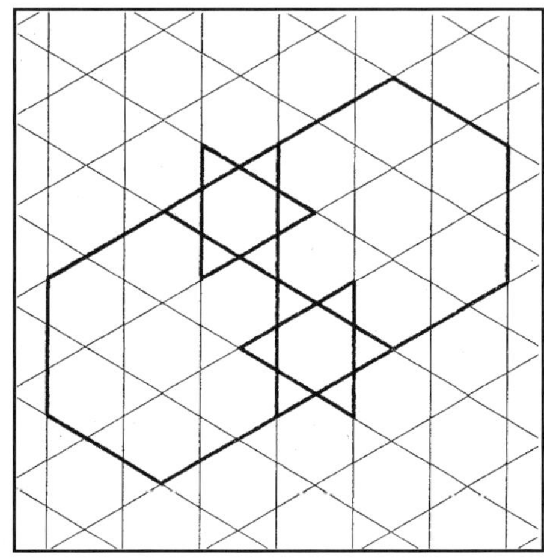

$6\ 2\ ^6/_2\ | \qquad tt^2\{6,^6/_2\}$

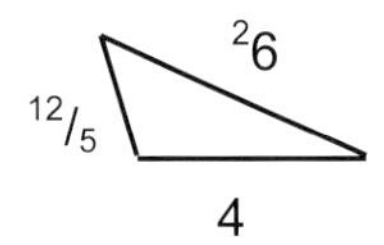

Analogue of Truncated
Dodecadodecahedron

Density 3 most apparent: every
double hexagon concentric with a
dodecagon

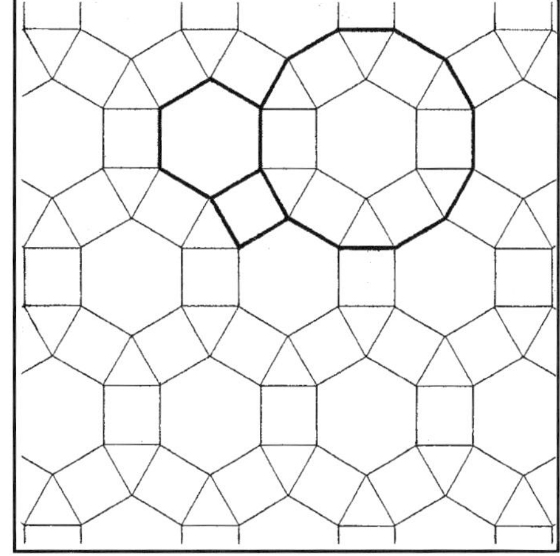

$^6/_5\ 2\ ^6/_2| \qquad tt^2\{6',^6/_2\}$

Analogue of Pentaquasitruncated
Dodecadodecahedron

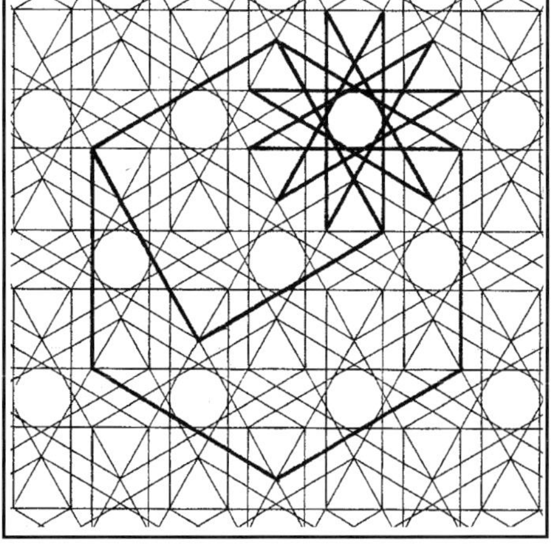

EVEN FACED TILINGS

All four even faced tilings shown below and opposite have one double polygon face within each triangular vertex figure. Interpreted correctly these double polygons produce ditrigonal vertex figures of type 2 with six polygons present at each vertex. See ($^5/_2$ 2 5) star polyhedral examples on pages 28 & 29 and later chapter on polyhedra with such vertices.

$^6/_4$ 2 6| $tt^2\{6,{}^6/_2{}'\}$

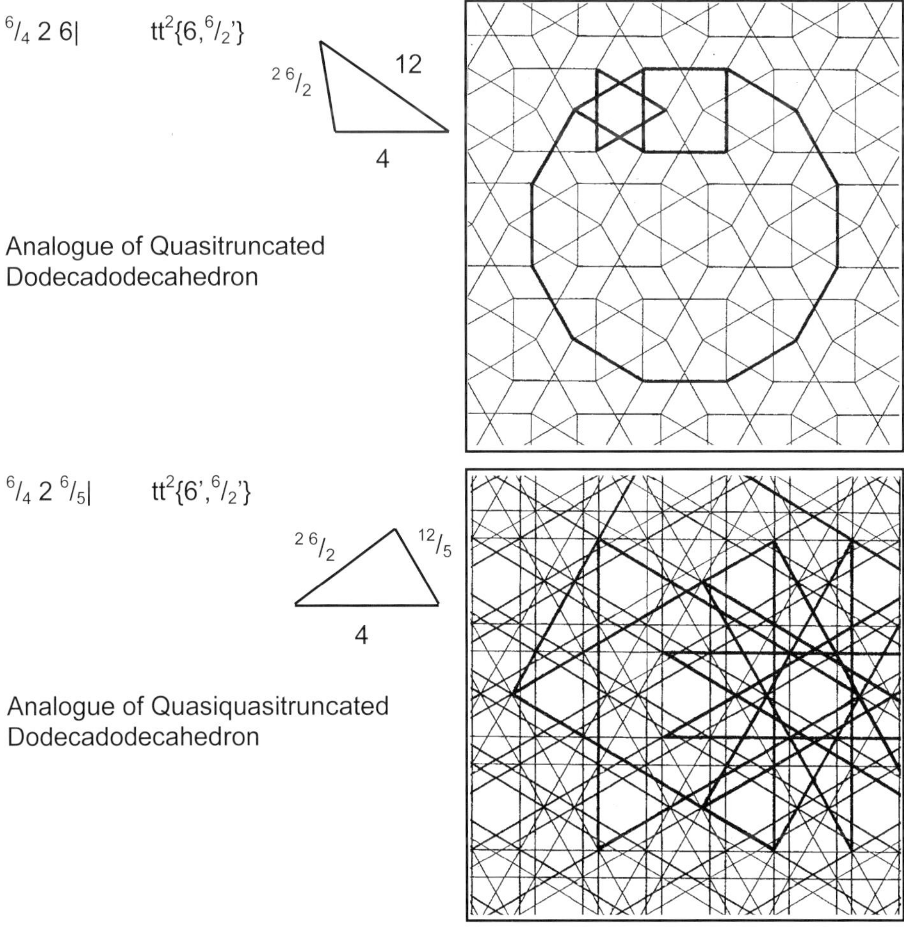

Analogue of Quasitruncated
Dodecadodecahedron

$^6/_4$ 2 $^6/_5$| $tt^2\{6',{}^6/_2{}'\}$

Analogue of Quasiquasitruncated
Dodecadodecahedron

THE ($^6/_2$ 2 6) FAMILY

$^6/_2$ 6 |2 $t^2t^2\{6,^6/_2\}$

Analogue of
Rhombidodecadodecahedron

|$^6/_2$ 2 6

Analogue of Snub
Dodecadodecahedron

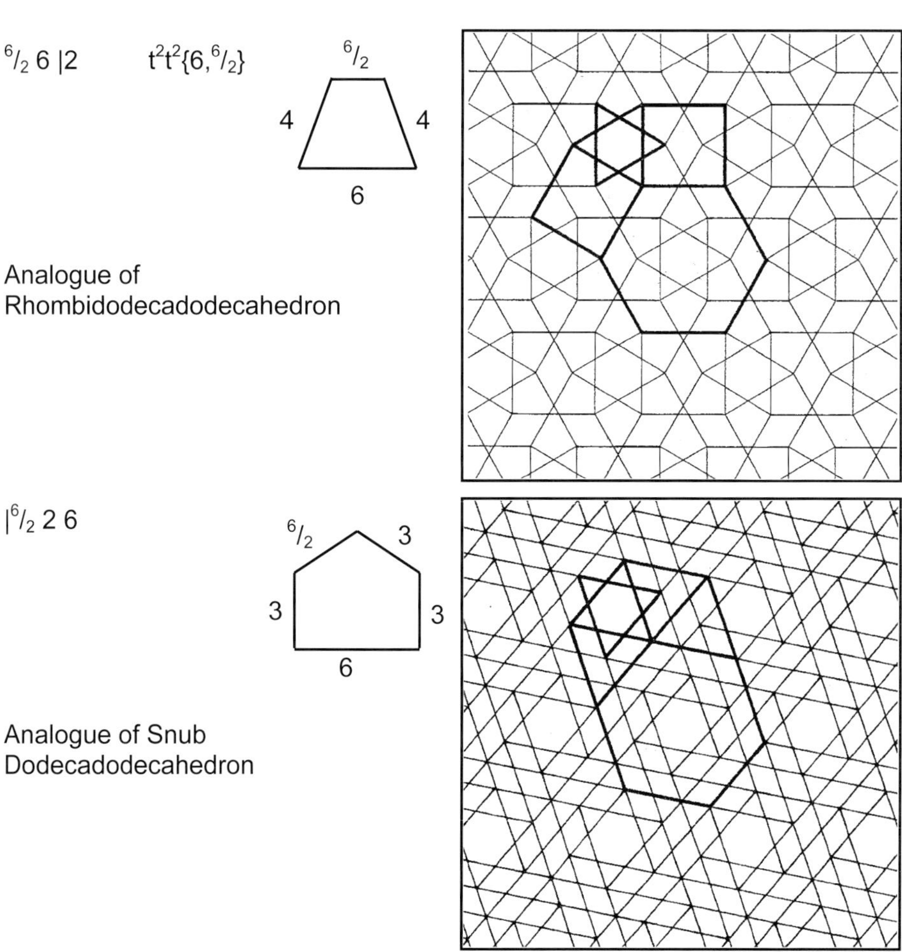

RHOMBI & SNUB TILINGS

Opposite and below are shown the rhombi and quasirhombi tilings for this family respectively, each with its associated snub form. The latter's vertex figures do not need to show the redundant digon {2} between the two adjacent triangles {3}, as this is simply the edge between those two triangles.

$^6/_4\ 6\ |2 \qquad t^2t^2\{6,{}^6/_2{}'\}$

Analogue of
Quasirhombidodecadodecahedron

$|{}^6/_4\ 2\ 6$

Analogue of Quasisnub
Dodecadodecahedron

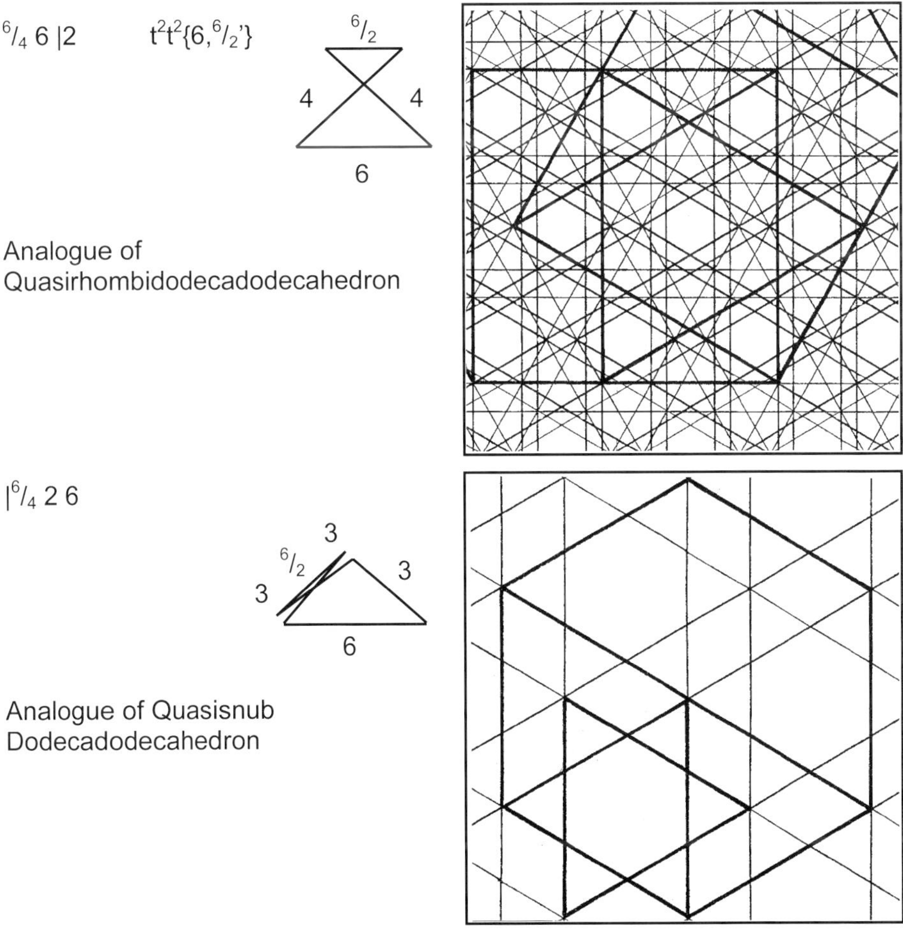

REFERENCES

Coxeter, H.S.M. 1973 *Regular Polytopes* Dover, New York

Coxeter, H.S.M., Longuet-Higgins, M.S. & Miller, J.C.P. 1953 *Uniform Polyhedra* Phil. Trans. R. Soc. Lond. A 246, 401-449

Critchlow, K. 1969 *Order in Space* Thames and Hudson

Grünbaum, B. & Shephard, G.C. 1989 *Tilings and Patterns; An Introduction* W.H.Freeman

Open University 1994 *Tilings* M336 Block 1, Unit IB1

Taylor, P. 1997 *The Complete? Polygon* Nattygrafix

Taylor, P. 1998 *Incomplete Tilings* Nattygrafix

In parallel with the star polyhedra and tilings of the previous two chapters, we now come to the consider the polyhedra made possible by creating two new Schwarz triangles from within the (3 2 4) net of Möbius triangles. They are the twofold (3 $^4/_2$ 3) and threefold ($^4/_2$ 2 4) Schwarz triangles and the results follow in pages set out to match those previously seen for the twofold (3 $^5/_2$ 3) and threefold ($^5/_2$ 2 5) Schwarz triangles of the star polyhedra, to which they are direct analogues.

They employ the polygon $\{^4/_2\}$ as described by the alternative notation adopted in this book, i.e this cross polygon appears as a simple cross comprising two out of phase digons. If the currently held view of $\{^4/_2\}$ were to be adopted we would be dealing with a double coincident digon and all the new polyhedra that might present themselves would simply be doubled up versions of the (3 2 4) polyhedra, which we already know about, again all rather boring.

The interpretation of these polygons in the alternative notation introduces an 'out of phase' element that allows these polygons to fill the spherical net in new and interesting ways, which are effectively kaleidoscopic. However it will be seen that the results are consistent and even in terms of covering the sphere twice or thrice over as applicable.

The essential contents of the following chapter were originally published in 1995 as Paper One of *Additions to the Uniform Polyhedra,* the material now re-presented with minor amendments. The abstract to that paper read:

The uniform polyhedra presented by Coxeter, Longuet-Higgins and Miller (1953) and confirmed by Skilling (1974) have essentially overlooked a small number of polyhedra, those in the ($^4/_2$ 2 4) and (3 $^4/_2$ 3) families. These result from accepting the star polygon $\{^4/_2\}$ as a face in its own right and are now described.

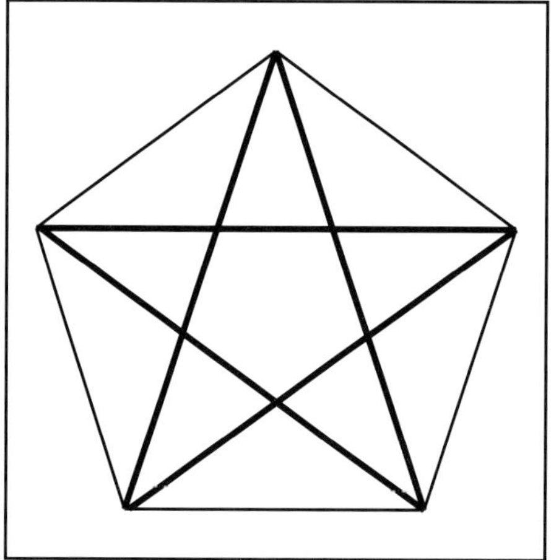

Inscribed Pentagon

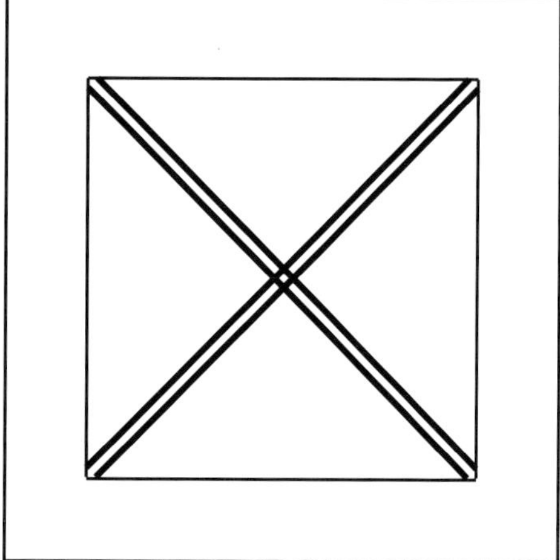

Inscribed Square

The cross polygon $\{^4/_2\}$ can be derived from a square $\{4\}$ in the same way that the star polygon $\{^5/_2\}$ is derived from a pentagon $\{5\}$. This can be achieved by either inscribing within or stellating without as shown above and below respectively.

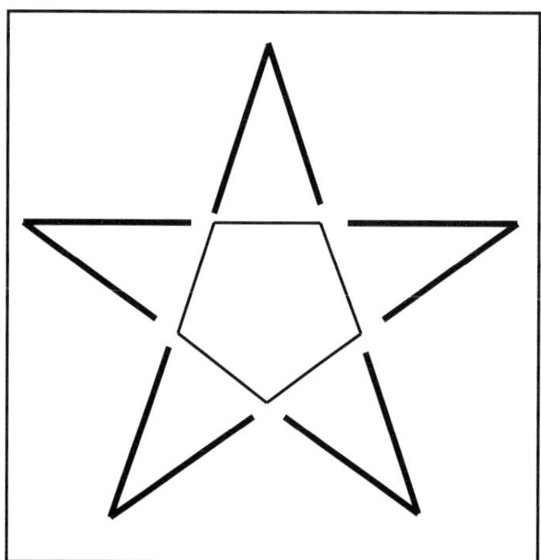

Stellated Pentagon

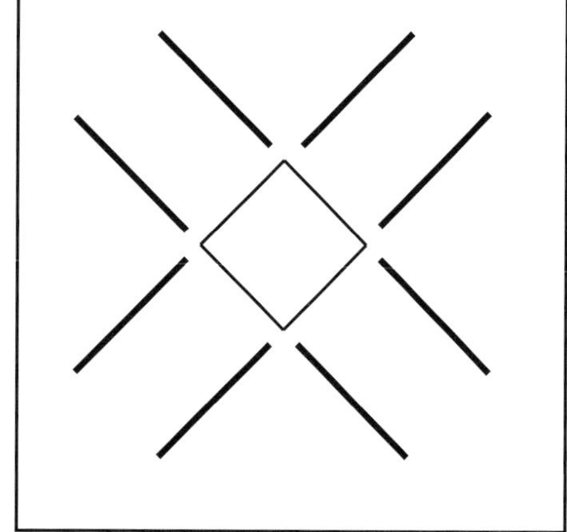

Stellated Square

In a similar way to the rationalisation of the polygon $\{2\}$ into a digon, the seemingly infinite stellation of the square $\{^4/_2\}$ can be seen as two such digons perpendicularly bisecting each other. The acceptance of this face into the polygonal vocabulary leads to two consistent families of polyhedra.

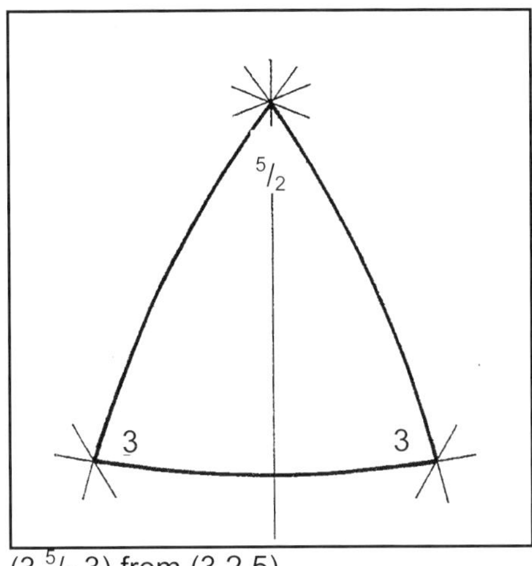

(3 $^5/_2$ 3) from (3 2 5)

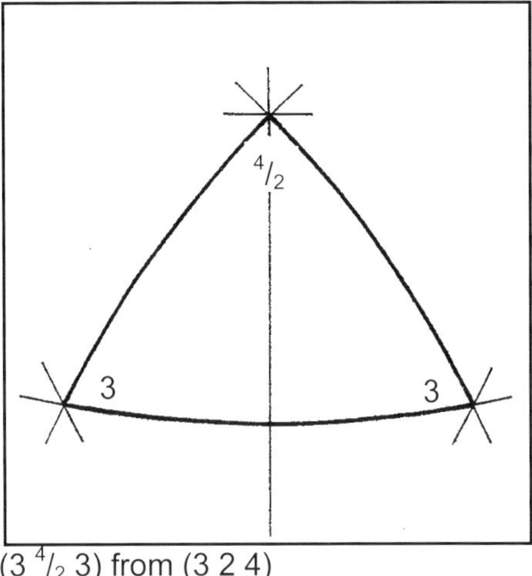

(3 $^4/_2$ 3) from (3 2 4)

Coxeter et al. have clearly shown how the (3 $^5/_2$ 3) Schwarz triangle can be derived from the (3 2 5) icosahedral Möbius triangle. This can be paralleled by the derivation of a (3 $^4/_2$ 3) Schwarz triangle from the (3 2 4) octahedral group, both families having density 2 as each Schwarz triangle contains two Möbius triangles.

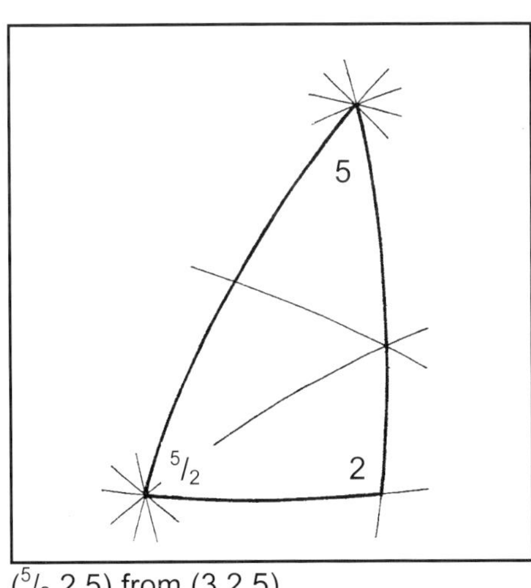

($^5/_2$ 2 5) from (3 2 5)

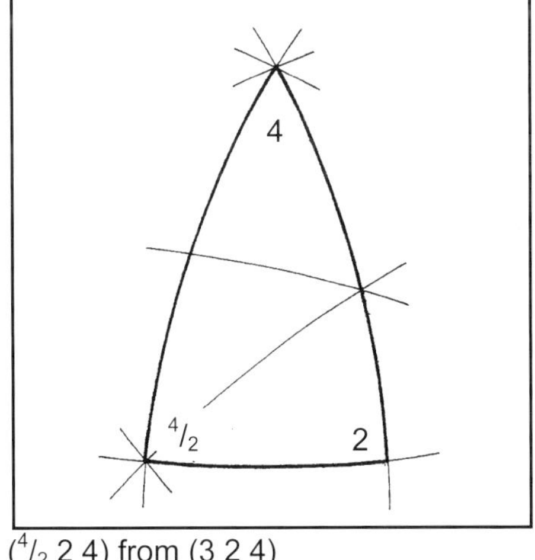

($^4/_2$ 2 4) from (3 2 4)

Similarly the derivation of the ($^5/_2$ 2 5) Schwarz triangle from the (3 2 5) icosahedral Möbius triangle can be paralleled by the derivation of a ($^4/_2$ 2 4) Schwarz triangle from the (3 2 4) octahedral Möbius triangle, both families having density 3, as each Schwarz triangle contains three Möbius triangles. We thus get a complete set of analogous inter-related polyhedra.

THE (3 $^4/_2$ 3) FAMILY

3| $^4/_2$ 3

Additions 1.4a

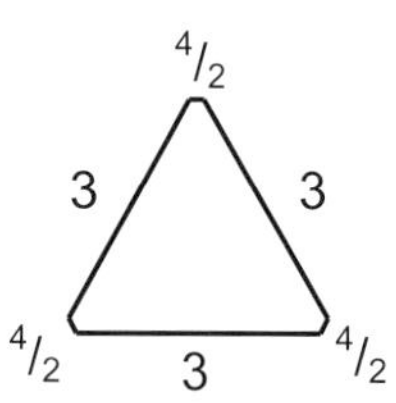

Stella Octangula

Analogue of Small Ditrigonal Icosidodecahedron 3| $^5/_2$ 3. Consists of two interpenetrating tetrahedra {3,3}. First described by Kepler but not to date regarded as a uniform polyhedron, rather a stellation

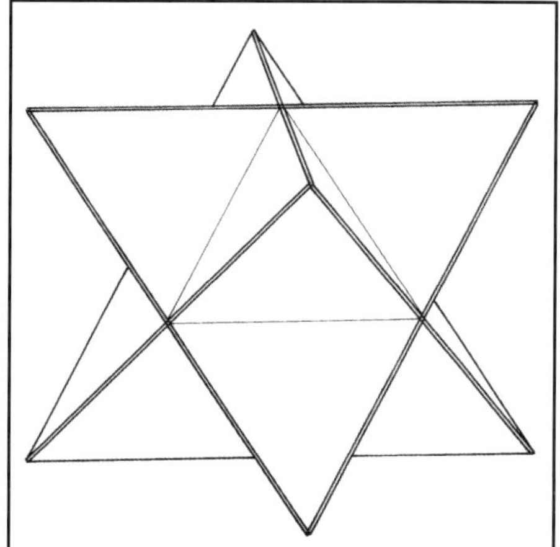

3 $^4/_2$ |3

Additions 1.4b

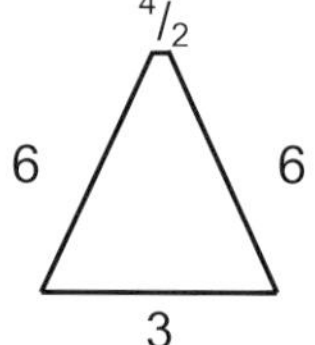

Truncated Stella Octangula

Analogue of Small Icosicosidodecahedron 3 $^5/_2$ |3. Consists of two interpenetrating truncated tetrahedra t{3,3}. Illustrated in Holden (1971) where it is dismissed because it decomposes

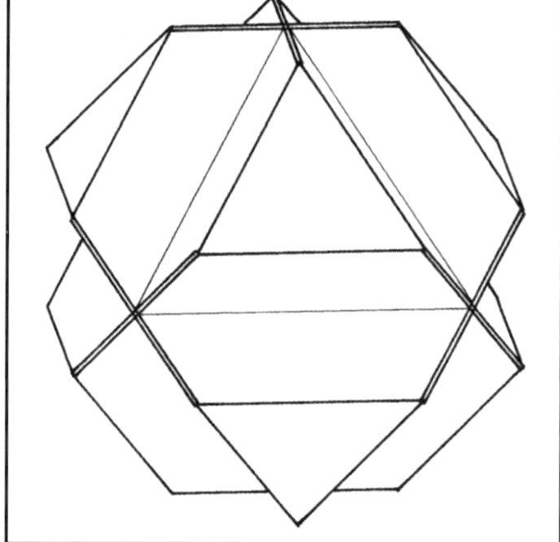

|3 $^4/_2$ 3

Additions 1.4c

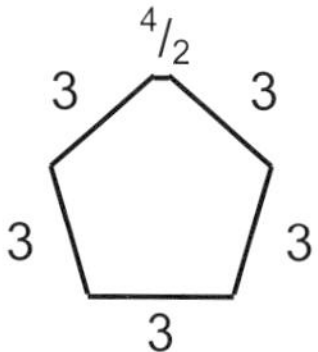

Snub Stella Octangula

Analogue of Small Snub Icosicosidodecahedron |3 $^5/_2$ 3. Consists of two interpenetrating snub tetrahedra (icosahedra) held together by the {$^4/_2$} cross polygons

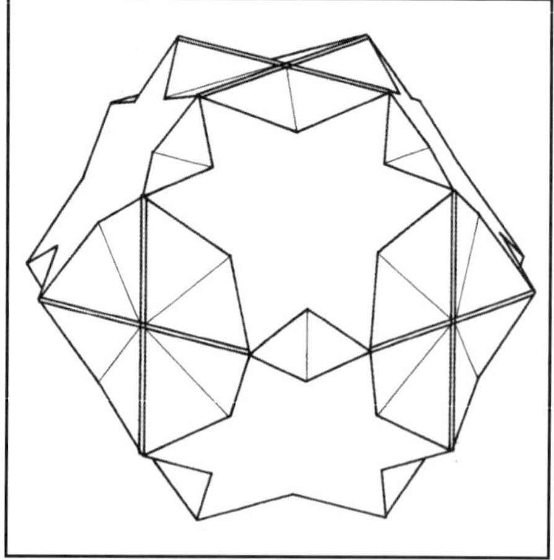

$3\ ^4/_2\ ^3/_2|$

Additions -

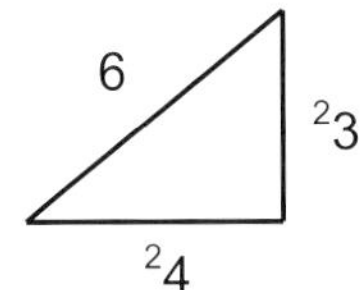

Double Cuboctahedron

Analogue of Quasiquasitruncated Small Ditrigonal Icosidodecahedron $3\ ^5/_3\ ^3/_2|$. Not included in *'Additions'* but later published in *'The Simpler? Polyhedra'* for completeness

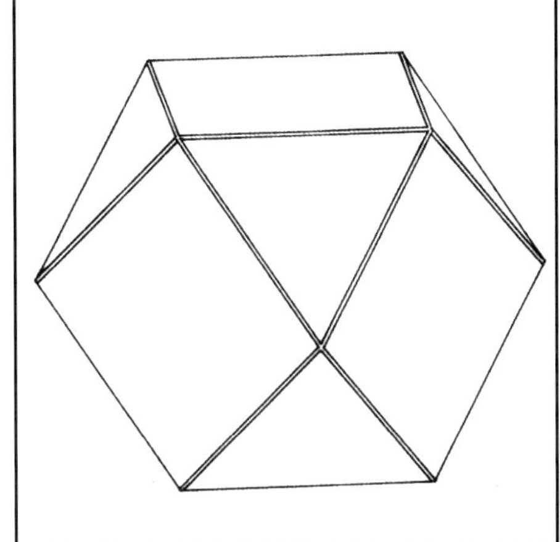

$3\ ^3/_2\ |^4/_2$

Additions -

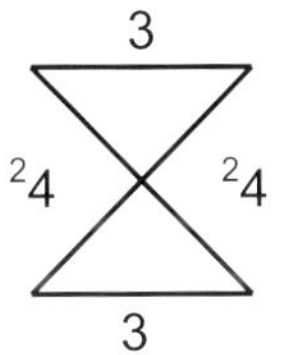

Double Tetrahemihexahedron

Analogue of Great Icosihemidodecahedron $3\ ^3/_2\ |^5/_3$. Not included in *'Additions'* but later published in *'The Simpler? Polyhedra'* for completeness

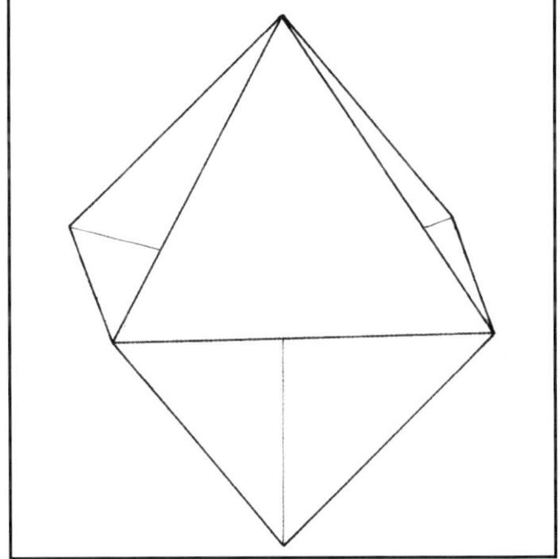

$|3\ ^4/_2\ ^3/_2$

Additions -

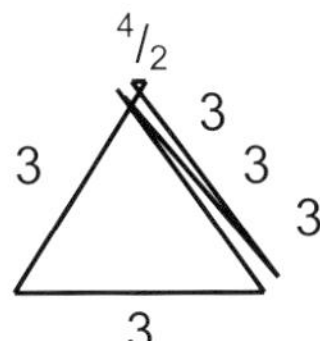

Quasisnub Stella Octangula

Analogue of Quasisnub Icosicosidodecahedron $|3\ ^5/_3\ ^3/_2$. Consists of two interpenetrating quasisnub tetrahedra held together by the $\{^4/_2\}$ cross polygons

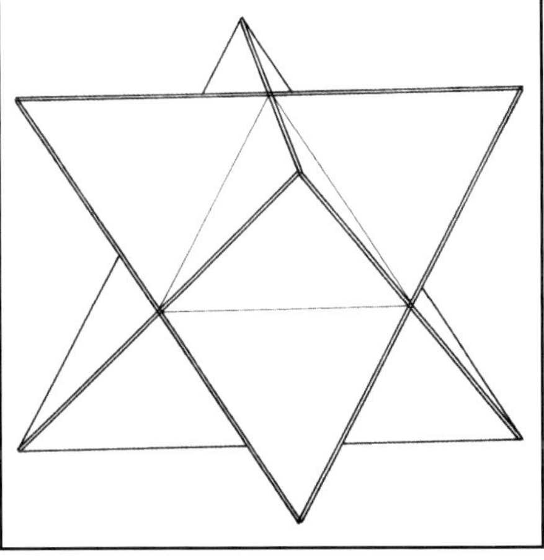

$3\ ^4/_2\ |^3/_2$

Additions -

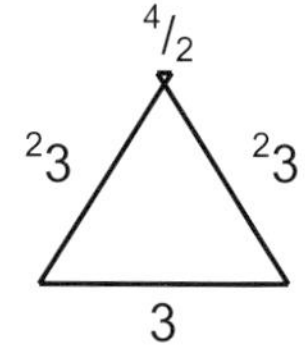

Double Stella Octangula

Analogue of Small Quasicosicosi-
dodecahedron $3\ ^5/_3\ |^3/_2$. Not included
in *'Additions'* but later published in
'The Simpler? Polyhedra' for
completeness

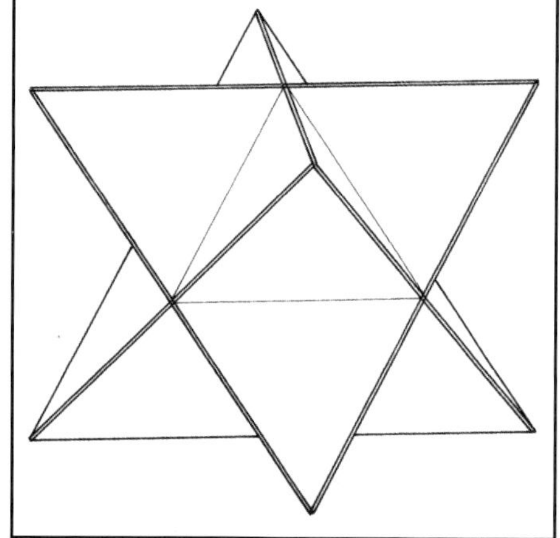

$|^3/_2\ ^4/_2\ ^3/_2$

Additions 1.4d

Retrosnub Stella Octangula

Analogue of Small Inverted
Retrosnub Icosicosidodecahedron
$|^3/_2\ ^5/_2\ ^3/_2$. Consists of two
interpenetrating retrosnub
tetrahedra (great icosahedra) held
together by the $\{^4/_2\}$ cross polygons

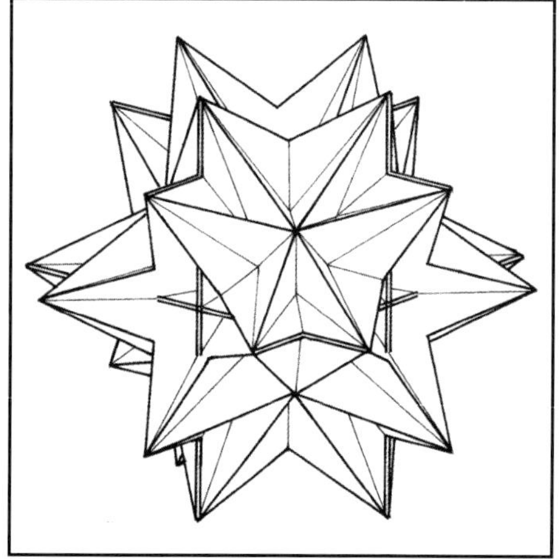

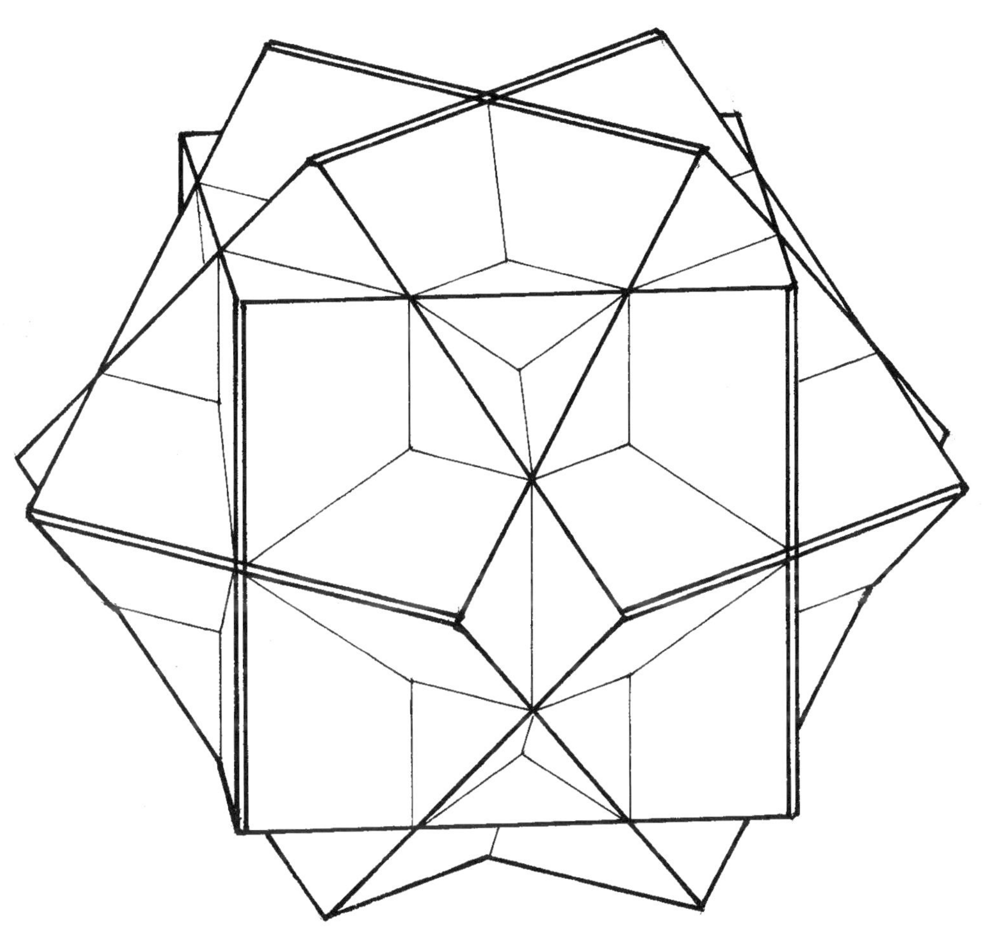

$^4/_2\,4\,|\,2 \qquad t^2t^2\{4,{}^4/_2\}$

THE ($^4/_2$ 2 4) FAMILY

$^4/_2$| 2 4   {4,$^4/_2$}

Additions 1.3a

φ_3
Great Hexahedron

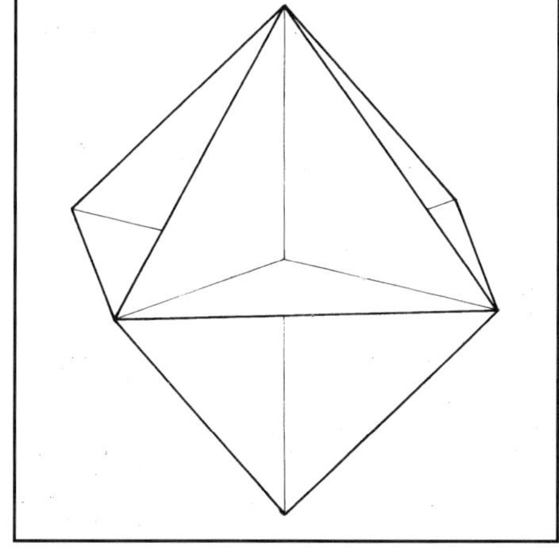

Analogue of Great Dodecahedron
$^5/_2$| 2 5   Consists of three mutually
perpendicular pairs of back to back
squares {4,2} inscribed within an
Octahedron {3,4}. Illustrated in
Holden (1971) as a 'regular nolid', a
polyhedron of zero volume.

$^4/_2$ 2 |4   t{4,$^4/_2$}

Additions 1.3b

Truncated Great Hexahedron

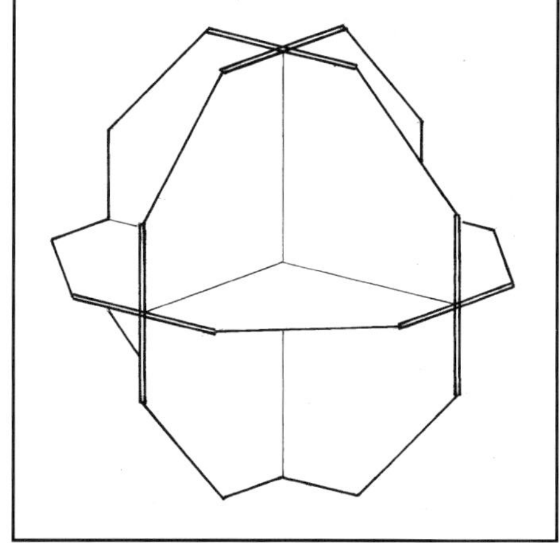

Analogue of Truncated Great
Dodecahedron $^5/_2$ 2 |5. Consists of
three mutually perpendicular pairs of
back to back octagons t{4,2}. This
and the quasitruncated version below
form a conjugate pair

$^4/_2$ 2 |$^4/_3$ t{4',$^4/_2$}

Additions 1.3c

Quasitruncated Great Hexahedron

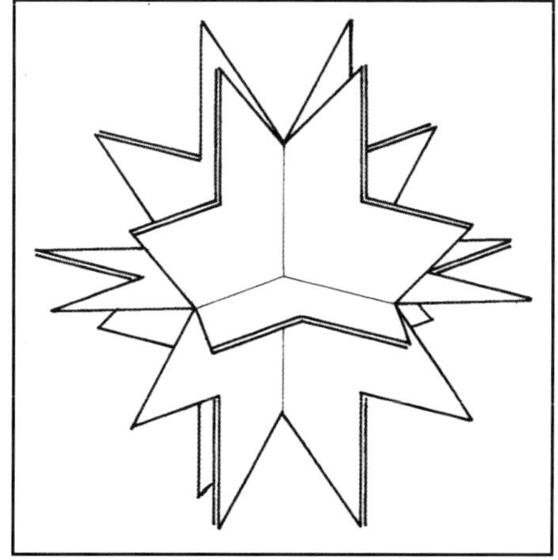

Analogue of Great Stellated
Dodecahedron which can be
regarded as a Quasitruncated Great
Dodecahedron $^5/_2$ 2 |$^5/_4$. Consists of
three mutually perpendicular pairs of
back to back octagrams t{4',2}

4| 2 $^4/_2$ {$^4/_2$,4}

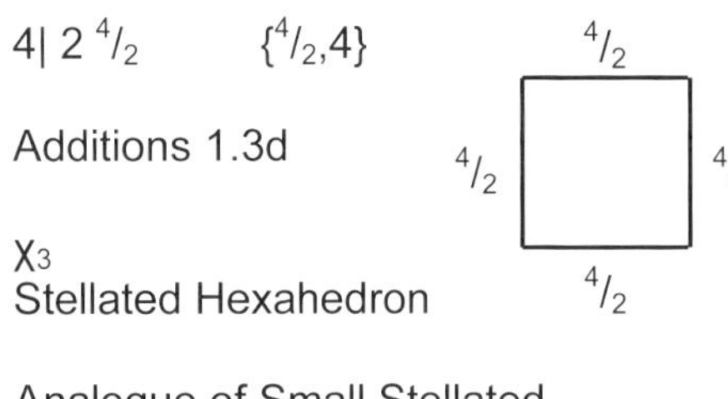

Additions 1.3d

Χ3
Stellated Hexahedron

Analogue of Small Stellated Dodecahedron 5| 2 $^5/_2$. Consists of three mutually perpendicular fourfold digons {2,4} stellating a Cube {4,3}. Illustrated in Wenninger (1983), where it is described as a Tetrahemihexacron

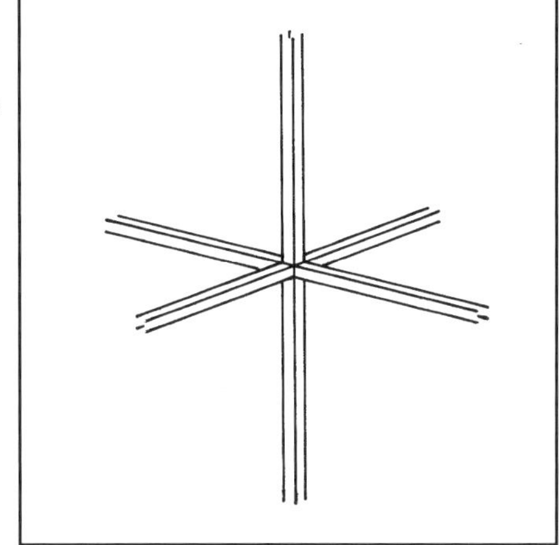

4 2 |$^4/_2$ t{$^4/_2$,4}

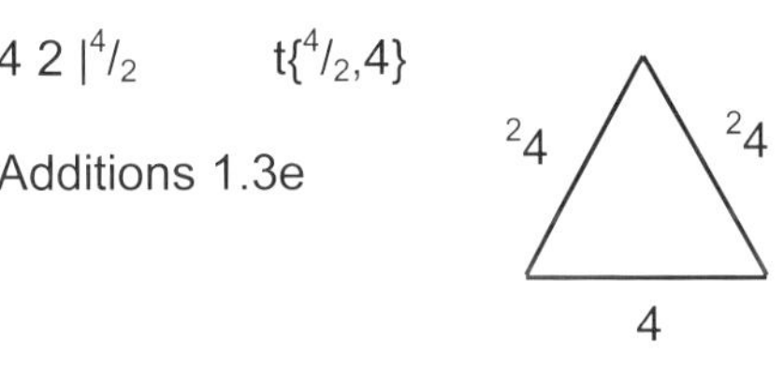

Additions 1.3e

Truncated Stellated Hexahedron

Analogue of Truncated Small Stellated Dodecahedron (three coincident Dodecahedra) 5 2 |$^5/_2$. Not included by Coxeter et al. (1953). Consists of three coincident Cubes with mutually perpendicular axes

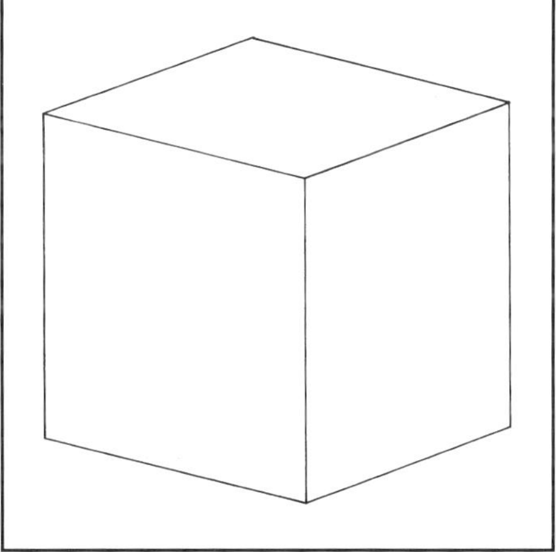

The Great Hexahedron can be seen as an inscribed Octahedron {3.4} and the Stellated Hexahedron is precisely that, a stellated Cube {4,3}.

The triple cube of the Truncated Stellated Hexahexahedron does not strictly qualify as a uniform polyhedron but has been left in for completeness and should appear in Coxeter et al.'s Table 6 of degenerate cases by adding in the value p = 4 (with density = 3) to the figure 2 p |$^p/_2$.

Because {$^4/_2$} is self dual it quasitruncates to produce the same double square that is produced by ordinary truncation so that we find that there is not a distinct Quasitruncated Stellated Hexahedron.

THE ($^4/_2$ 2 4) FAMILY

2| $^4/_2$ 4   t^2\{4,$^4/_2$\}

Additions 1.3f

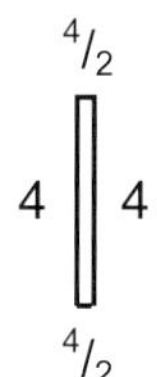

Hexahexahedron

Analogue of Dodecadodecahedron
2| $^5/_2$ 5. Consists of three mutually
perpendicular pairs of back to back
squares. Illustrated in Holden (1971)
where it is another 'regular nolid', a
polyhedron of zero volume

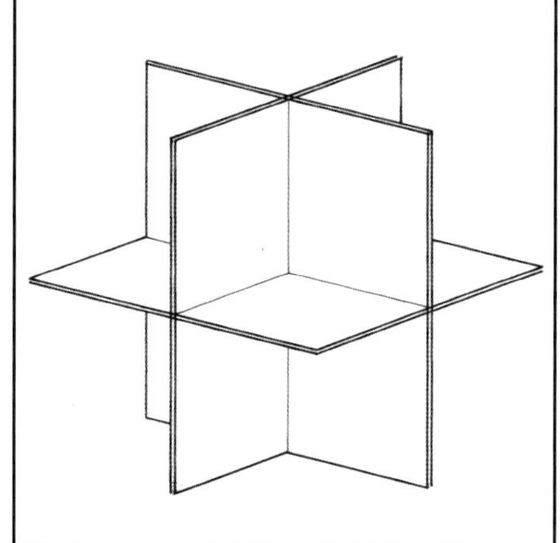

4 2 $^4/_2$| tt^2\{4,$^4/_2$\}

Additions 1.3g

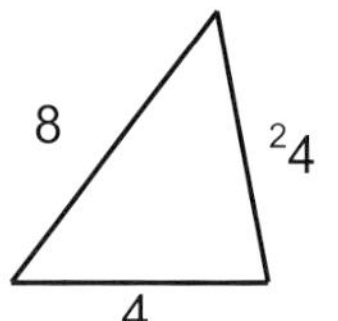

Truncated Hexahexahedron

Analogue of Truncated
Dodecadodecahedron 5 2 $^5/_2$|, not
included by Coxeter et al. (1953).
Consists of three mutually
perpendicular octagon prisms. One
of a conjugate pair

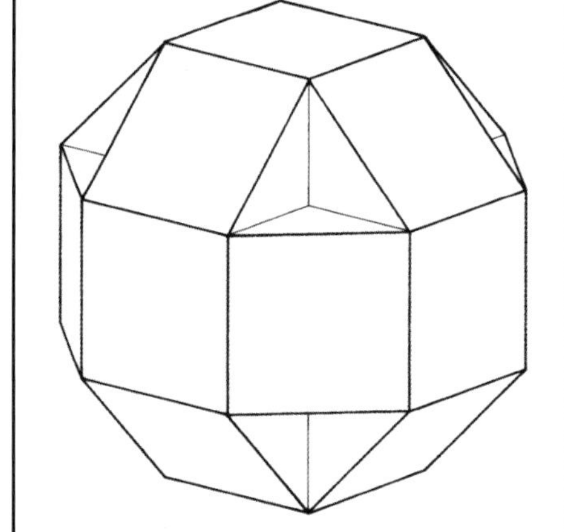

$^4/_3$ 2 $^4/_2$| tt^2\{4',$^4/_2$\}

Additions 1.3h

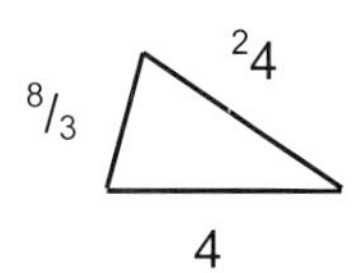

Quasitruncated Hexahexahedron

Analogue of Quasitruncated
Dodecadodecahedron $^5/_4$ 2 $^5/_2$|, not
included by Coxeter et al. (1953).
Consists of three mutually
perpendicular octagram prisms.
Illustrated in Holden (1971)

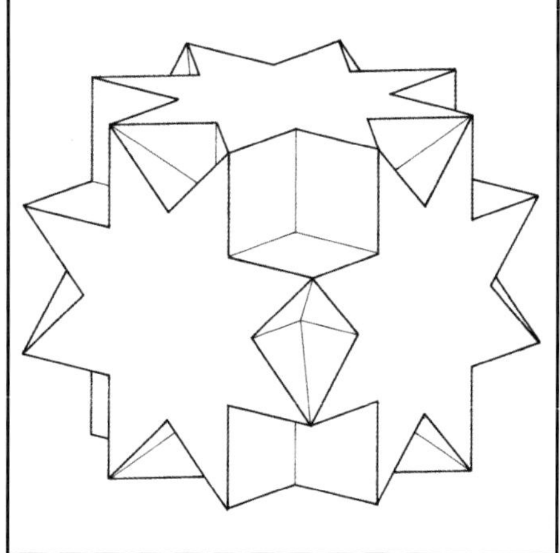

Again because $\{^4/_2\}$ is self dual there are no further quasitruncations of the quasi-regular hexahexahedron $t^2\{4,^4/_2\}$. There are neither a Quasitruncated Hexahexahedron (where $\{^4/_2\}$ would be the quasitruncated face) nor a Quasiquasitruncated Hexahexahedron.   The   pair   of   even   faced truncations   shown   opposite   are   all   that   we   get   and   like   their   star polyhedra analogues they involve type 2 ditrigonal vertex figures (see later chapter p.83).

Also since $\{^4/_2\}$ has zero edge length in a vertex figure, if given a twist it is still $\{^4/_2\}$, so there is not a distinct Quasirhombihexahexahedron, nor a Quasisnub Hexahexahedron derived from it. All that we get of these forms are the two polyhedra shown below.

$^4/_2$ 4 |2 $t^2t^2\{4,^4/_2\}$

Additions 1.3i

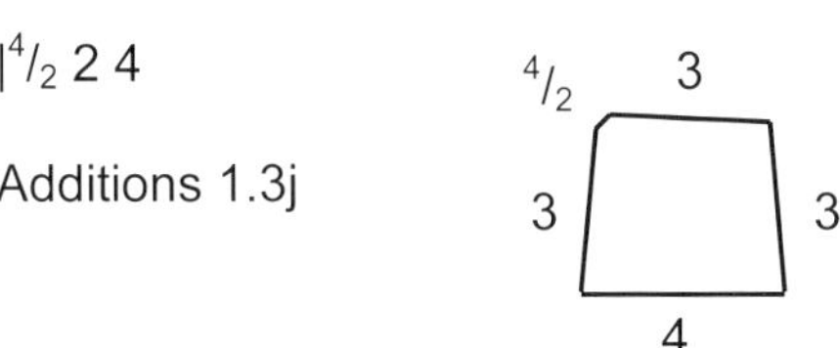

Rhombihexahexahedron

Analogue of Rhombidodecadodeca-
hedron $^5/_2$ 5 |2. Consists of three
mutually perpendicular
interpenetrating cubes. Illustrated in
Holden (1971). Appears in M C
Escher picture 'Waterfall'

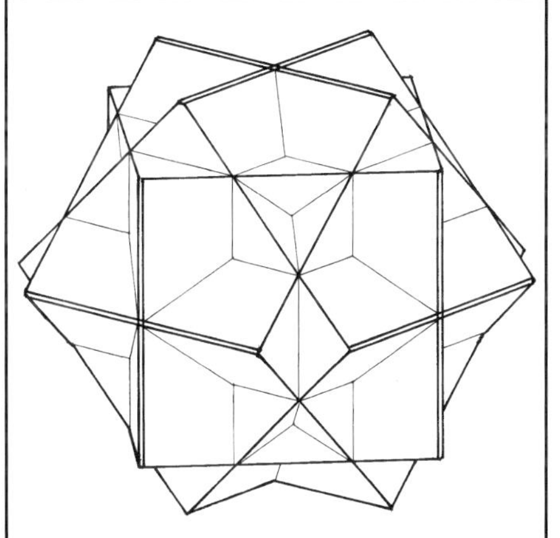

|$^4/_2$ 2 4

Additions 1.3j

Snub Hexahexahedron

Analogue of Snub Dodecadodeca-
hedron |$^5/_2$ 2 5. Consists of three
mutually perpendicular
interpenetrating square antiprisms.
One of an enantiomorphous pair with
a left or right handed twist

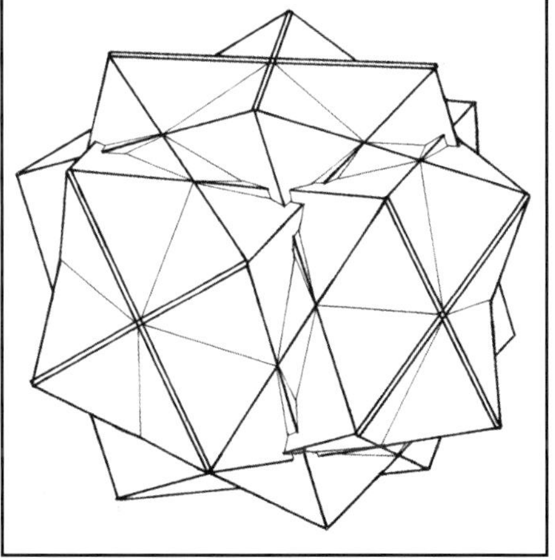

CONCLUSION

It has always seemed strange that Kepler's Stella Octangula, despite oozing with symmetry and often illustrating the stellation of an octahedron, has to date escaped classification amongst the uniform polyhedra. The acceptance of the $\{^4/_2\}$ face into the polygonal vocabulary achieves this and also ties in a host of other figures previously relegated to the miscellany.

Amongst these are some of the 'regular nolids' illustrated by Holden (1971), together with the present author's $^4/_2$ 2 |4 and $^4/_2$ 2 |$^4/_3$ and Wenninger's Tetrahemihexacron, all of which might still be discounted as having no volume. These are however analogous to dihedral members of the (2 2 q) families from which Coxeter et al. derive the prisms and antiprisms. These last properly considered include digons amongst their faces, so that the inclusion here of a crossed digon $\{^4/_2\}$ is but a small further step.

The remaining figures are all solids of finite volume although comprised of two or three simpler solids compounded. If you accept the $\{^4/_2\}$ faces as described by the alternative notation, which join the simpler solids together, all of these figures do have equal numbers of the same regular polygons at every vertex and therefore qualify as uniform polyhedra. Several of these have appeared without classification in Holden (1971) but the author believes five of them have never before been fully described and can conveniently be considered in two groups:

Firstly there are the 4 2 $^4/_2$| and $^4/_3$ 2 $^4/_2$| which form a conjugate pair and consist of three mutually perpendicular octagon and octagram prisms respectively. These like their analogues in the ($^5/_2$ 2 5) family would be dismissed by Coxeter et al. (1953) for having three faces meeting at some edges. As will be explained in the later chapter on ditrigonal vertex figures, because of the double polygons they actually have four faces at these edges and should be included for completeness.

Secondly there are the three new snubs, |$^4/_2$ 2 4, |3 $^4/_2$ 3 and |$^3/_2$ $^4/_2$ $^3/_2$, composed respectively of three square antiprisms, two icosahedra and two great icosahedra, linked by $\{^4/_2\}$ faces. Full details of the intersecting faces of these more complex figures are to be found in Appendix 1 opposite.

The set of figures described here include many that might have been dismissed for a variety of different reasons. However the author is of the opinion that because they form a discrete and complete set of interrelated forms, all derived from the two new Schwarz triangles, they deserve their place amongst the polyhedra proper.

Comparison with their analogues in the (3 $^5/_2$ 3) and ($^5/_2$ 2 5) families of star polyhedra in an earlier chapter should show enough similarities to convince.

Descriptions of the faces of the new snubs, dimensioned relative to an edge length 2:

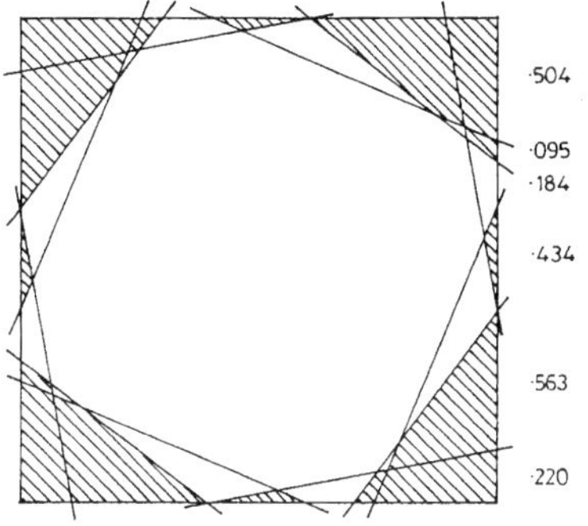

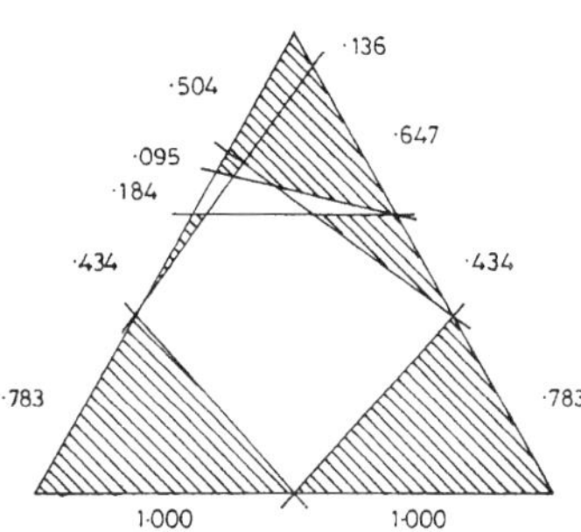

Snub Hexahexahedron |$^4/_2$ 2 4

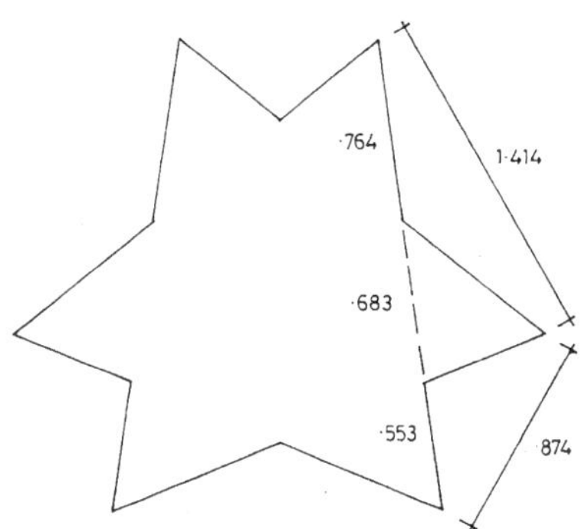

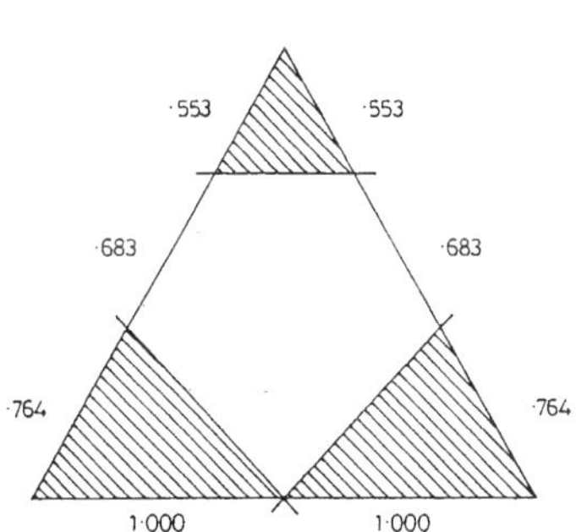

Snub Stella Octangula |3 $^4/_2$ 3

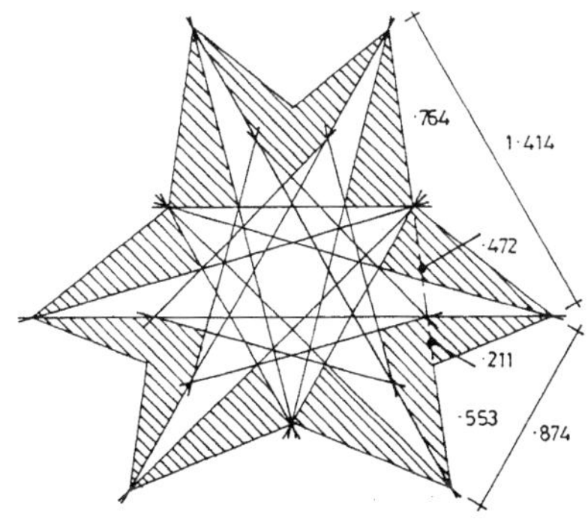

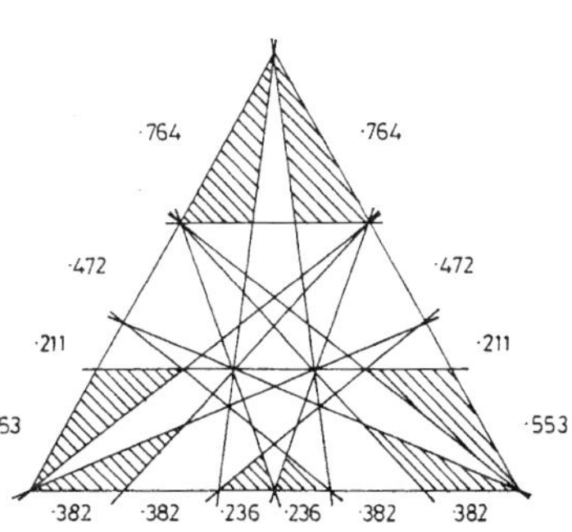

Retrosnub Stella Octangula |$^3/_2$ $^4/_2$ $^3/_2$

59

REFERENCES

Coxeter, H.S.M. 1973 *Regular Polytopes* Dover, New York

Coxeter, H.S.M., Longuet-Higgins, M.S. & Miller, J.C.P. 1953 *Uniform Polyhedra* Phil. Trans. R. Soc. Lond. A 246, 401-449

Critchlow, K. 1969 *Order in Space* Thames and Hudson

Cundy, H.M. & Rollett, A.P. 1951 *Mathematical Models* Oxford U P

Escher, M.C. 1992 *The Graphic Work* Benedikt Taschen

Holden, A. 1971 *Shapes, Space, and Symmetry* Columbia U P

Skilling, J. 1974 *The Complete Set of Uniform Polyhedra* Phil. Trans. R. Soc. Lond. A 278, 111-135

Taylor, P. 1995 *Additions to the Uniform Polyhedra* Nattygrafix

Taylor, P. 1997 *The Complete? Polygon* Nattygrafix

Taylor, P. 1998 *Incomplete Tilings* Nattygrafix

Taylor, P. 1999 *The Simpler? Polyhedra* Nattygrafix

Taylor, P. 2000 *The Star & Cross Polyhedra* Nattygrafix

Wenninger, M.J. 1971 *Polyhedron Models* Cambridge U P

Wenninger, M.J. 1983 *Dual Models* Cambridge U P

In this next chapter we look at the results of introducing two of our new cross polygons $\{^4/_2\}$ and $\{^6/_3\}$ into the realm of the tilings. We will investigate what becomes possible when we effectively substitute $^4/_2$ for 2 and $^6/_3$ for 2 into our original (4 2 4) and (3 2 6) Möbius triangles respectively.

We thus find the twofold or fourfold (4 $^4/_2$ 4) and the fourfold (3 $^6/_3$ 6), two new Schwarz triangles for which the results follow in pages that can still provide parallels with those previously seen for the twofold (3 $^5/_2$ 3) and threefold ($^5/_2$ 2 5) Schwarz triangles of the star polyhedra.

As before they employ the polygons $\{^4/_2\}$ and $\{^6/_3\}$ as described by the alternative notation adopted in this book, i.e. these cross polygons appear as simple forms comprising two or three out of phase digons.

If the currently held view of these were to be adopted we would be dealing with double or triple coincident digons and all the new tilings that might present themselves would simply be multiple versions of the (4 2 4) or (3 2 6) tilings respectively, which again we already know about.

The interpretation of these polygons in the alternative notation introduces an 'out of phase' element that allows these tilings to fill the plane in these new and interesting ways, which are effectively kaleidoscopic. However it will be seen that the results are consistent and even in terms of filling the plane exactly twice or four times over as applicable.

We have seen that the tilings like the polyhedra have similar nets of Möbius triangles, (4 2 4) and (3 2 6), both of which are not spherical, but which stretch in the plane to infinity like the tilings themselves. In a similar way to that used to create the star polyhedra from the regular ones, the Möbius triangles of the basic tilings (4 2 4) and (3 2 6) can be taken in multiples to form larger Schwarz triangles such as those already seen with the star tilings.

Because the net of Mobius triangles extends to infinity rather than closing up on itself, we find that a (4 $^4/_2$ 4) Schwarz triangle can in fact be made comprising 2, 4, 8, 16, 32 etc. Möbius triangles, i.e. any power of 2 of them. Similarly a basic (4 2 4) can be found comprising 1, 9, 25, 49 etc. Möbius triangles, i.e. any odd square of them.

In this particular section we will look at a double version of the (4 2 4) Möbius triangle, where two {4} vertices become a {$^4/_2$} and a quadruple version of the (3 2 6) Möbius triangle where three {6} vertices form a {$^6/_3$}. The right angles within the Schwarz triangles for these are formed of two 4's and three 6's of the corresponding Möbius triangles respectively, giving rise to {$^4/_2$} and {$^6/_3$} as the symmetries there.

This exactly mirrors what happens when {$^5/_2$} is formed from two 5's in the icosahedral spherical net of Möbius triangles in the polyhedra to form the star polyhedra and seems a good reason to keep {$^4/_2$} and {$^6/_3$} as the notation for these cross polygons rather than 2{2} and 3{2} as some would have them labelled.

This relationship could be expressed in the currently accepted notation, but seems clumsy as it would involve two {4} vertices becoming a 2{2} or three {6} vertices becoming a 3{2}, as against the simpler relationship of two {5} vertices becoming a {$^5/_2$} that we have seen in the star polyhedra.

These larger Schwarz triangles can similarly be reflected around to fill the plane with points on their vertices, edges or faces forming new vertices for the various tilings and truncated tilings shown. In general if the new Schwarz triangle comprises k Möbius triangles, the new grid will cover the plane k times, with density k.

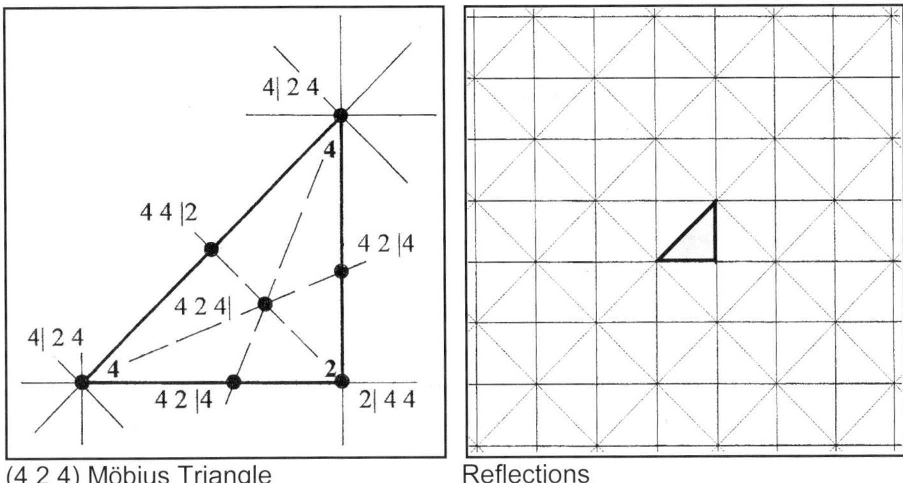

(4 2 4) Möbius Triangle   Reflections

The tilings of the {4,4} symmetry family can be generated from the (4 2 4) Möbius triangle, which by reflection along its edges can be made to tile the entire plane. Points on the vertices, midway along the edges and central to that triangle become the respective vertices of the regular/quasi-regular, the truncated/rhombi and the even faced tilings in that family.

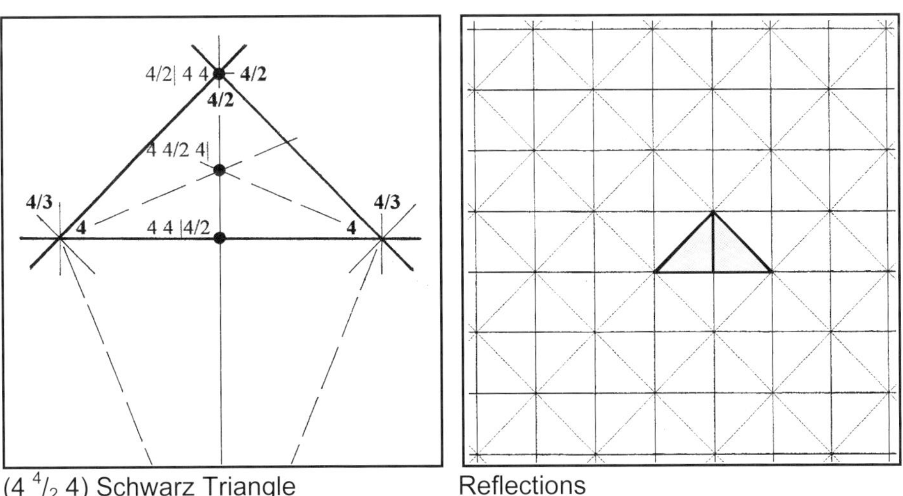

(4 $^4/_2$ 4) Schwarz Triangle   Reflections

Taking two of these Möbius triangles together we can form a Schwarz triangle (4 $^4/_2$ 4), which can similarly tile the plane by reflection and produce its own family of corresponding tilings. These will include {$^4/_2$} cross polygon faces or those of its truncation ^{2}{4}. Those with vertices on the axis of the Schwarz triangle will simply be double versions of certain (4 2 4) tilings.

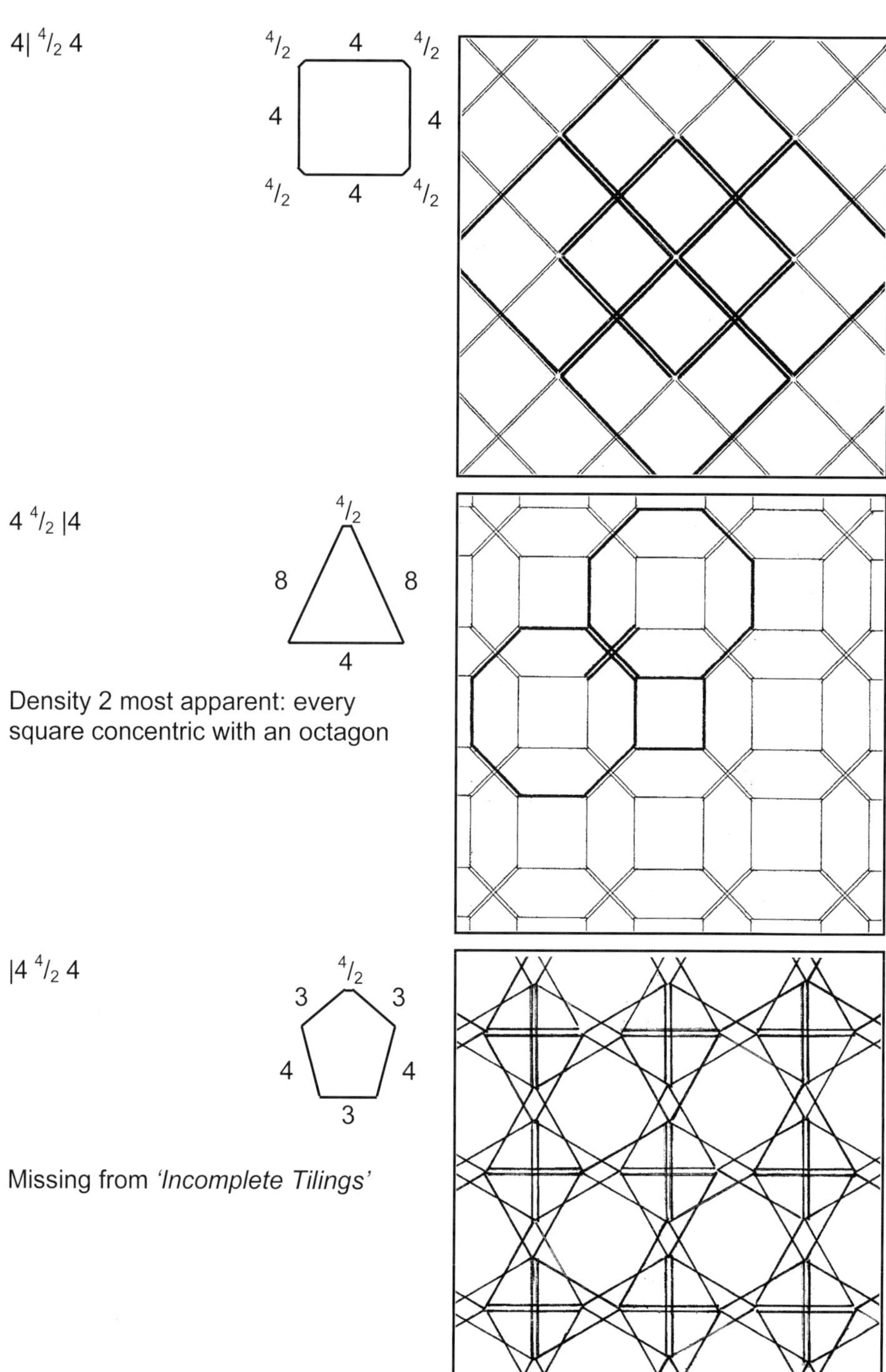

4| $^4/_2$ 4

4 $^4/_2$ |4

Density 2 most apparent: every
square concentric with an octagon

|4 $^4/_2$ 4

Missing from *'Incomplete Tilings'*

4 $^4/_2$ $^4/_3$|

4 $^4/_2$ |$^4/_3$

|$^4/_3$ $^4/_2$ $^4/_3$

Missing from *'Incomplete Tilings'*

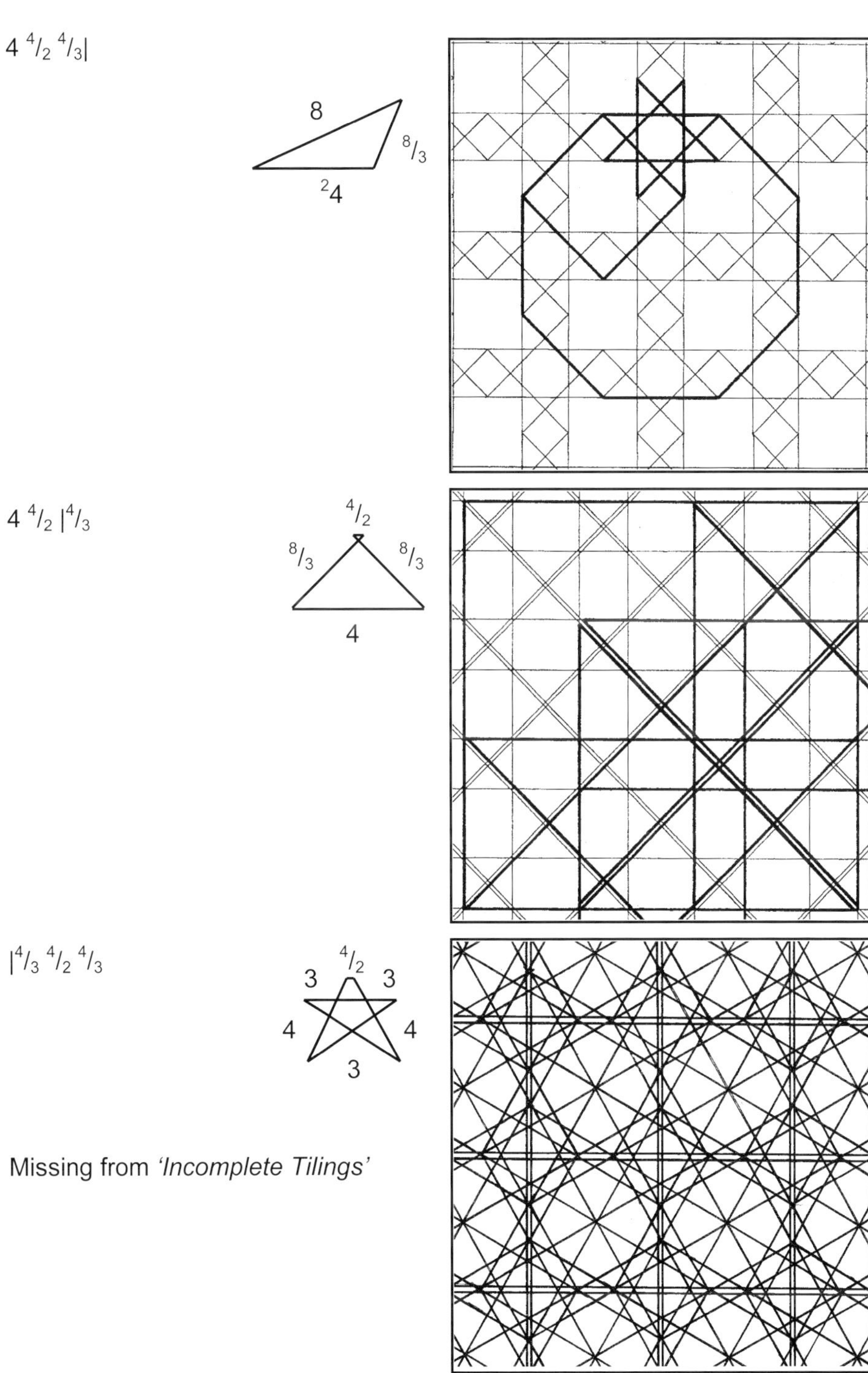

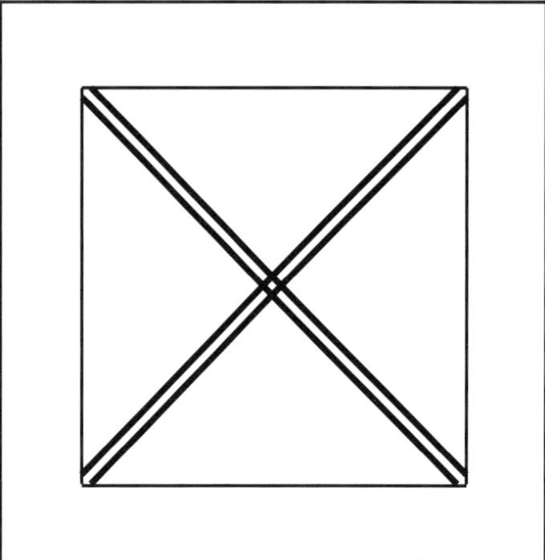

Inscribed Square

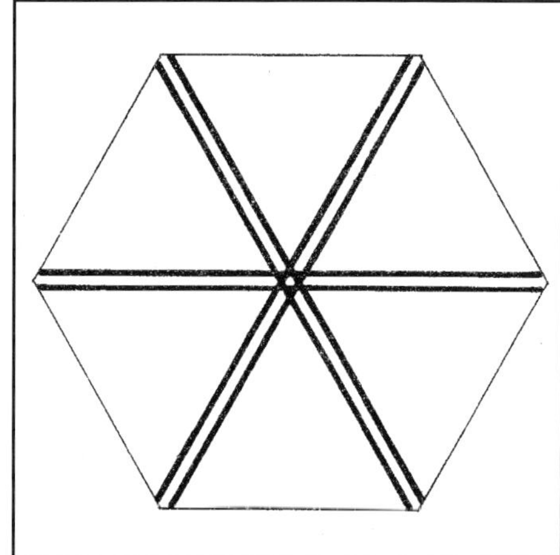

Inscribed Hexagon

The cross polygon $\{^6/_3\}$ can be derived from a hexagon $\{6\}$ in the same way that the cross polygon $\{^4/_2\}$ is derived from a square $\{4\}$. This can be achieved by either inscribing within or stellating without as shown above and below respectively.

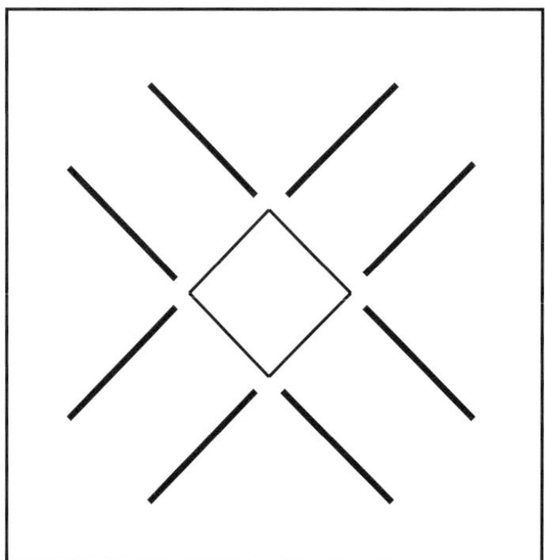

Stellated Square

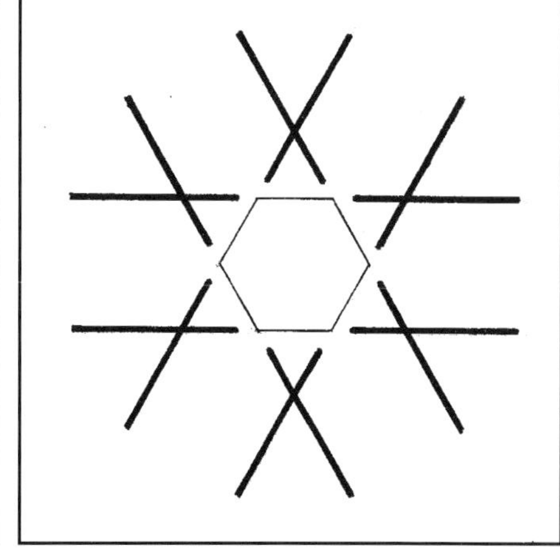

Stellated Hexagon

In a similar way to the rationalisation of the polygon $\{2\}$ into a digon, the seemingly infinite stellation of the hexagon $\{^6/_3\}$ can be seen as three such digons bisecting each other. The acceptance of this face into the polygonal vocabulary leads to a consistent family of tilings as follows.

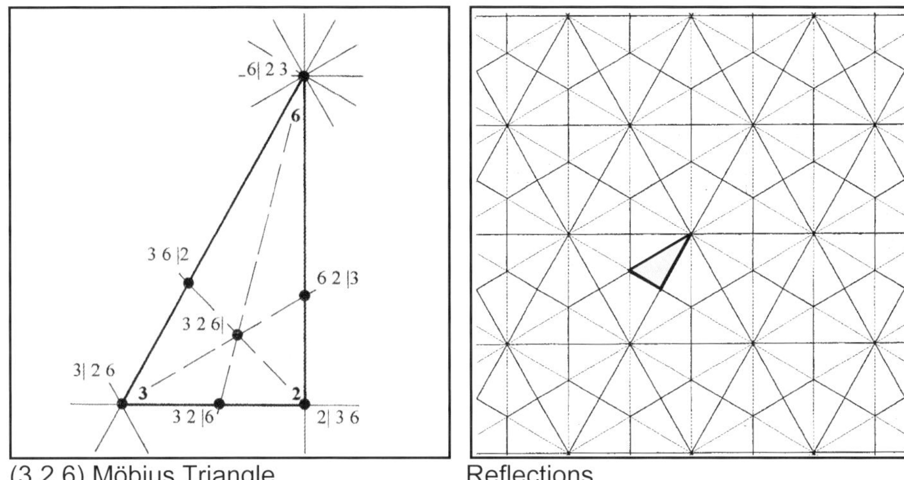

(3 2 6) Möbius Triangle   Reflections

The tilings of the {3,6} symmetry family can be generated from the (3 2 6) Möbius triangle, which by reflection along its edges can be made to tile the entire plane. Points on the vertices, midway along the edges and central to that triangle become the respective vertices of the regular/quasi-regular, the truncated/rhombi and the even faced tilings in that family.

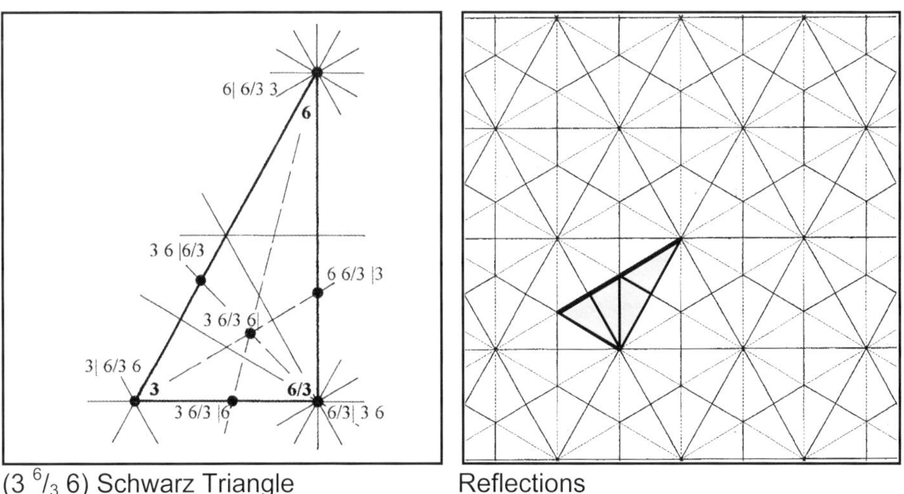

(3 $^6/_3$ 6) Schwarz Triangle   Reflections

Taking four of these Möbius triangles together we can form a Schwarz triangle (3 $^6/_3$ 6), which can similarly tile the plane by reflection and produce its own family of corresponding tilings, some of which will include {$^6/_3$} cross polygon faces. These are shown on the following pages and generally include {$^6/_3$} cross polygon faces, or those of its truncation {$^{12}/_3$}.

3| $^6/_3$ 6

3 $^6/_3$ |6

3 $^6/_3$ |$^6/_5$

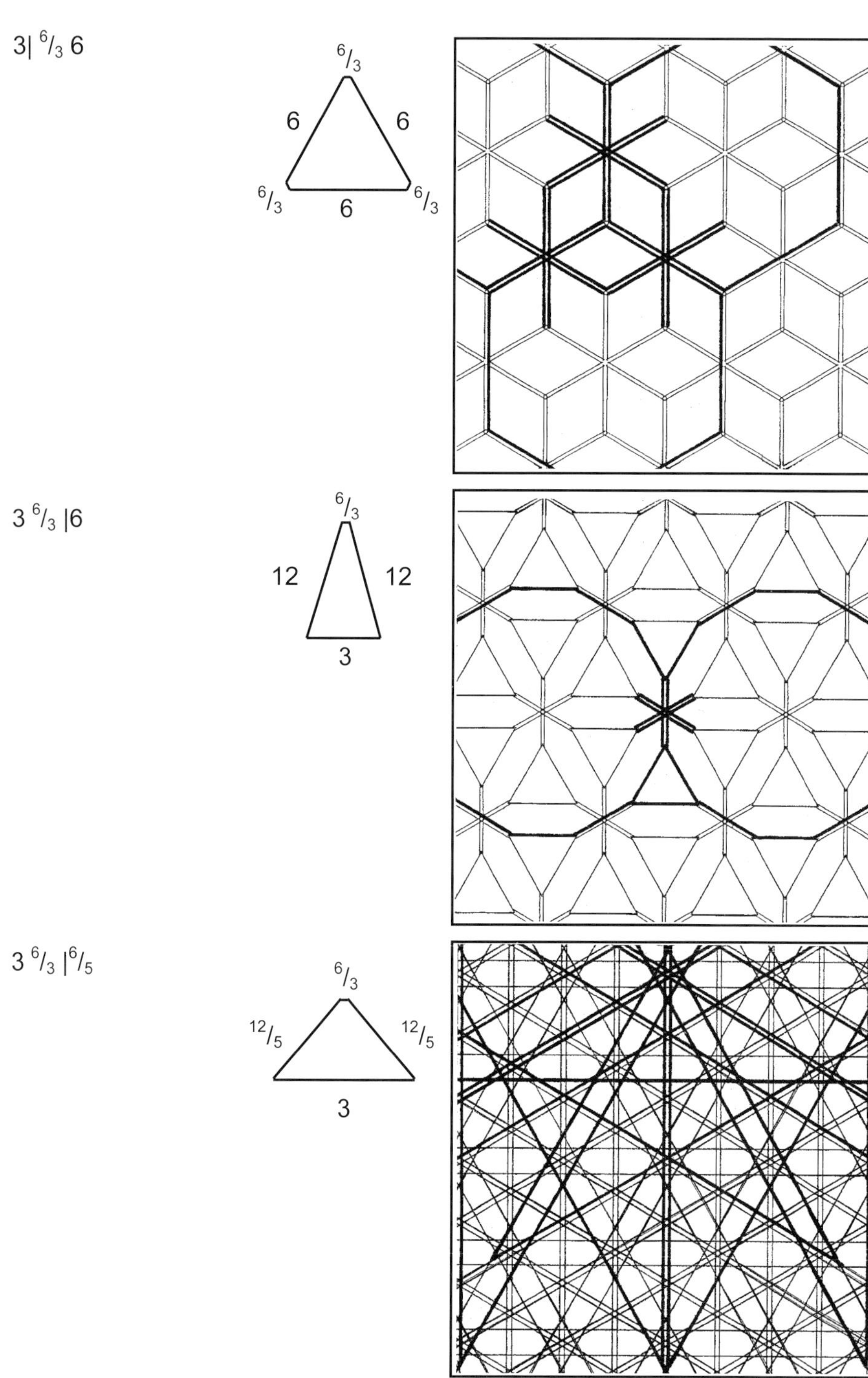

6| $^6/_3$ 3

6 $^6/_3$ |3

6 $^6/_3$ |$^3/_2$

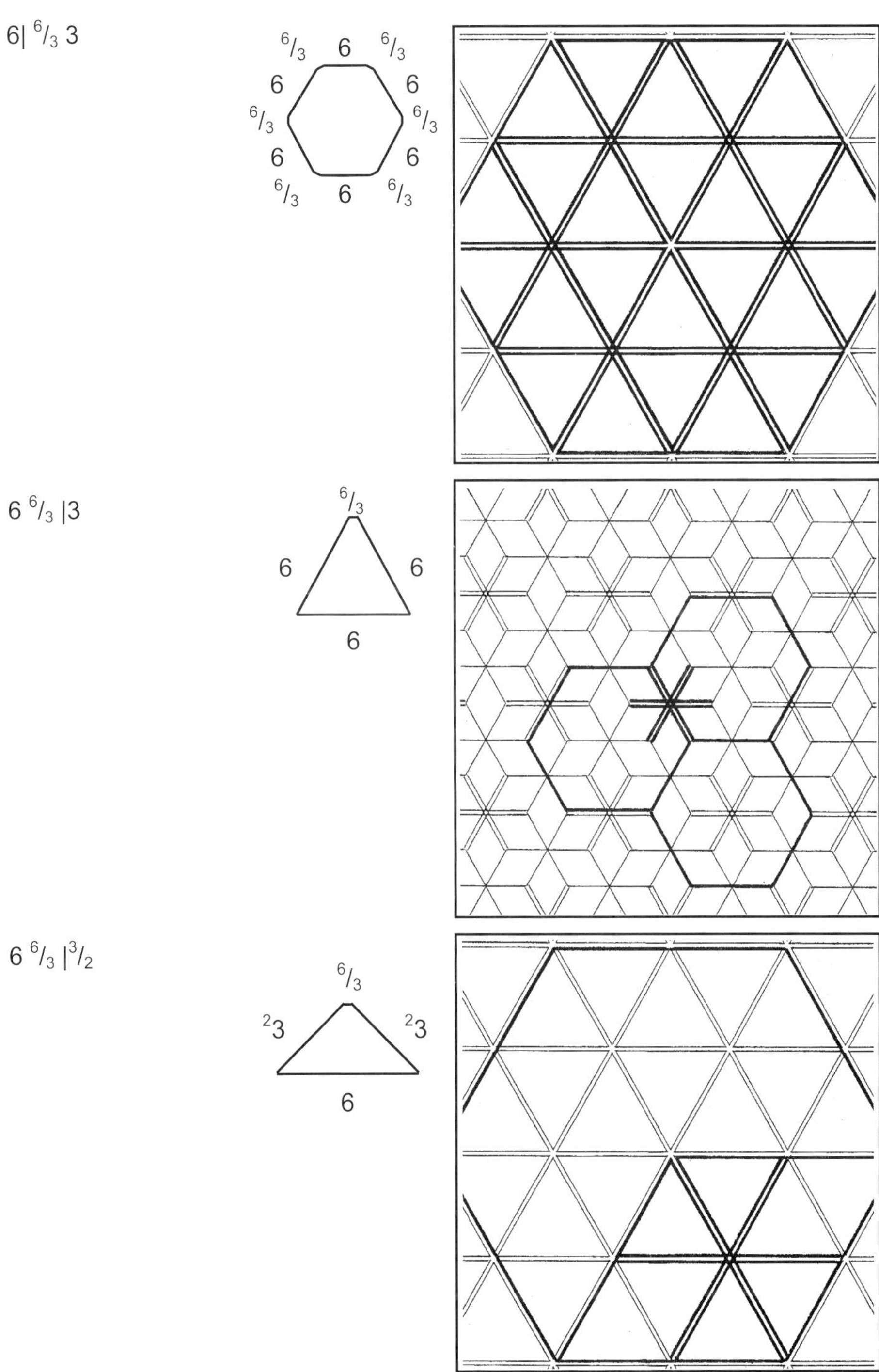

$^6/_3$| 3 6

3 $^6/_3$ 6|

Density 4 most apparent: every
dodecagram (three squares)
concentric with a dodecagon

3 $^6/_3$ $^6/_5$|

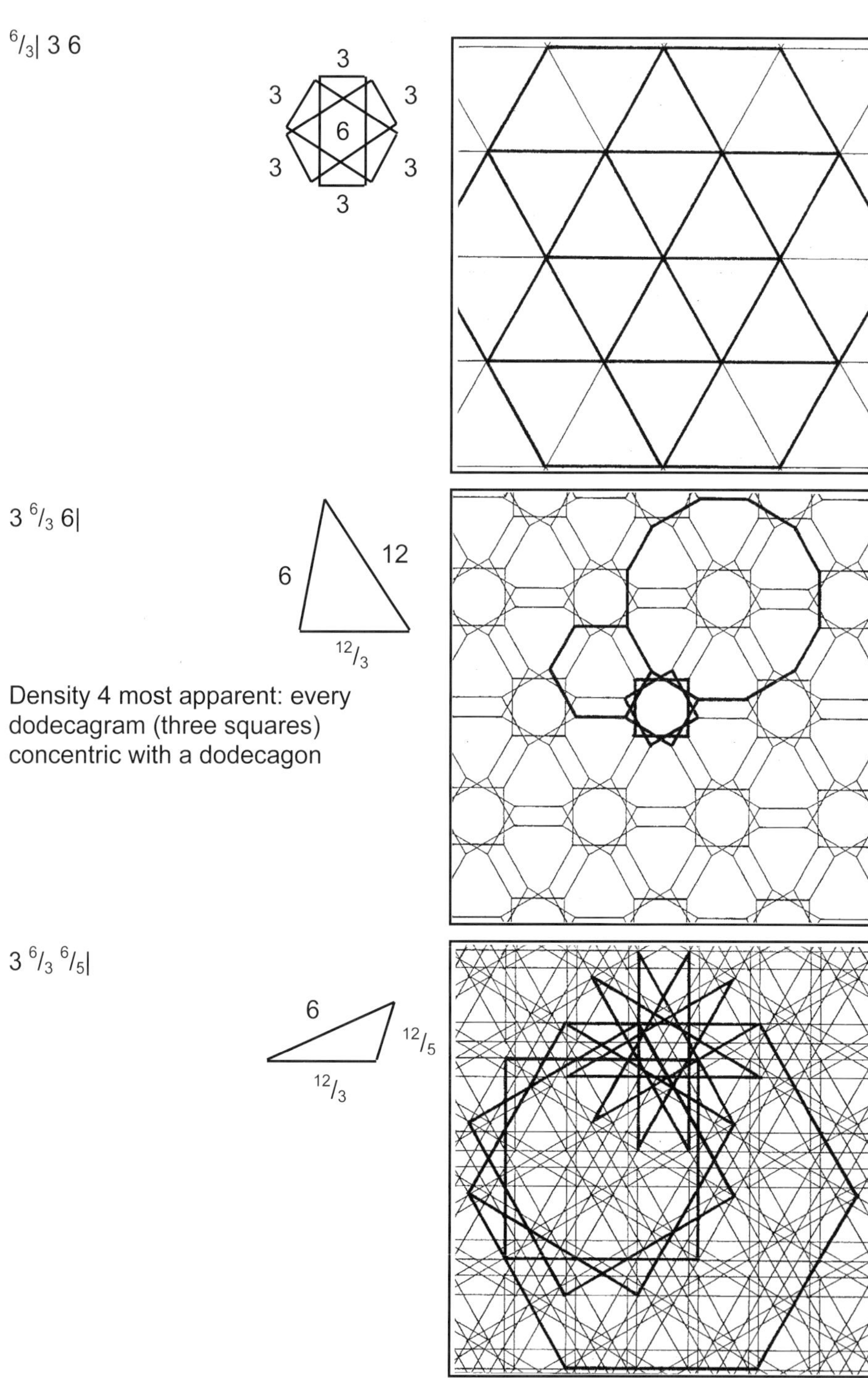

^{3}/$_2$ ^{6}/$_3$ 6|

^{3}/$_2$ ^{6}/$_3$ ^{6}/$_5$|

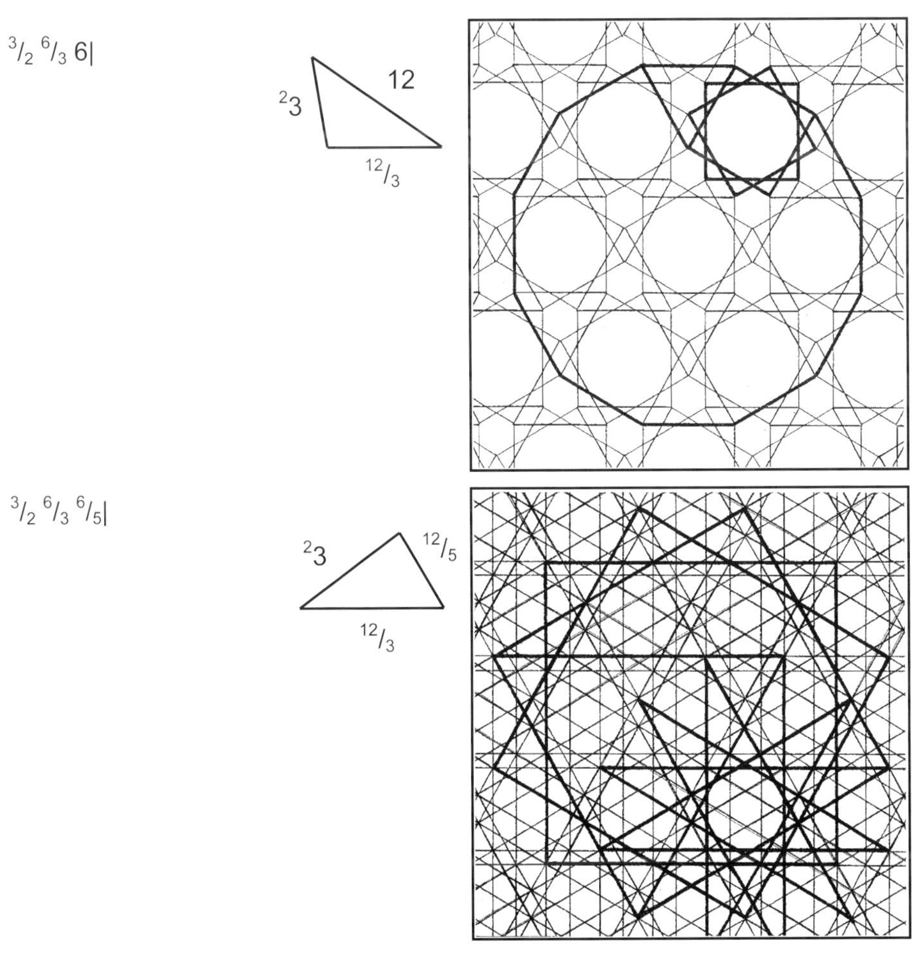

3 6 | $^6/_3$

|3 $^6/_3$ 6

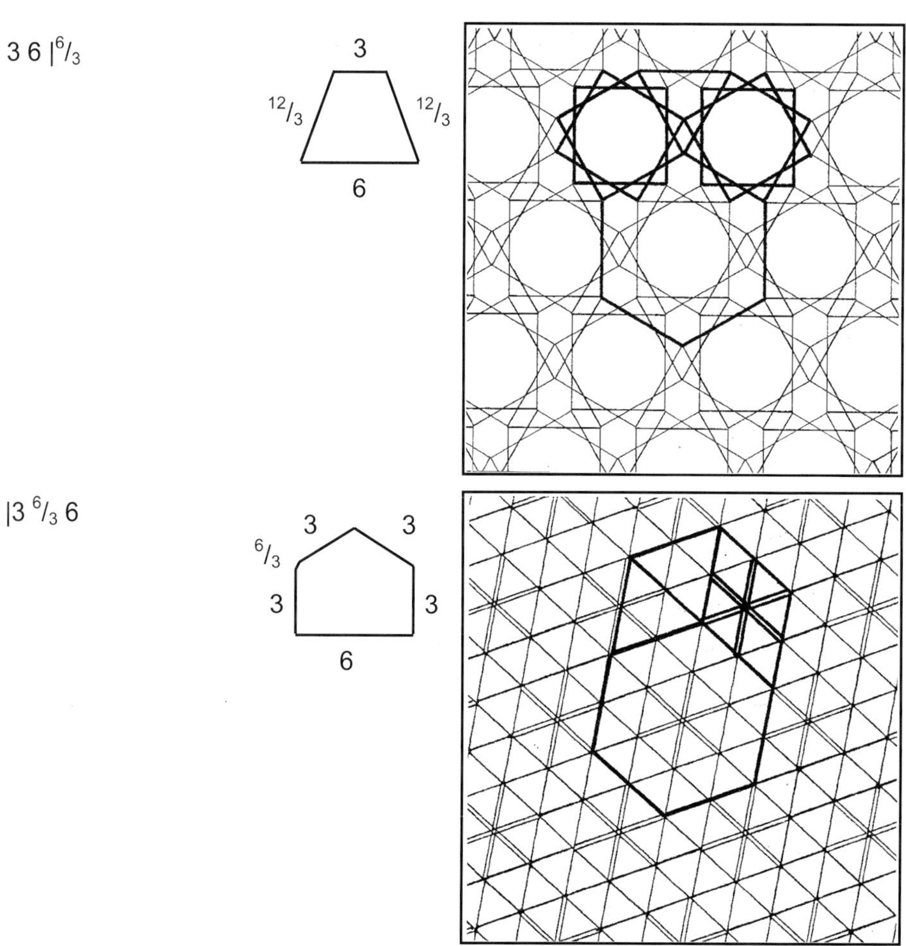

$^3/_2$ 6 | $^6/_3$

| $^3/_2$ $^6/_3$ 6

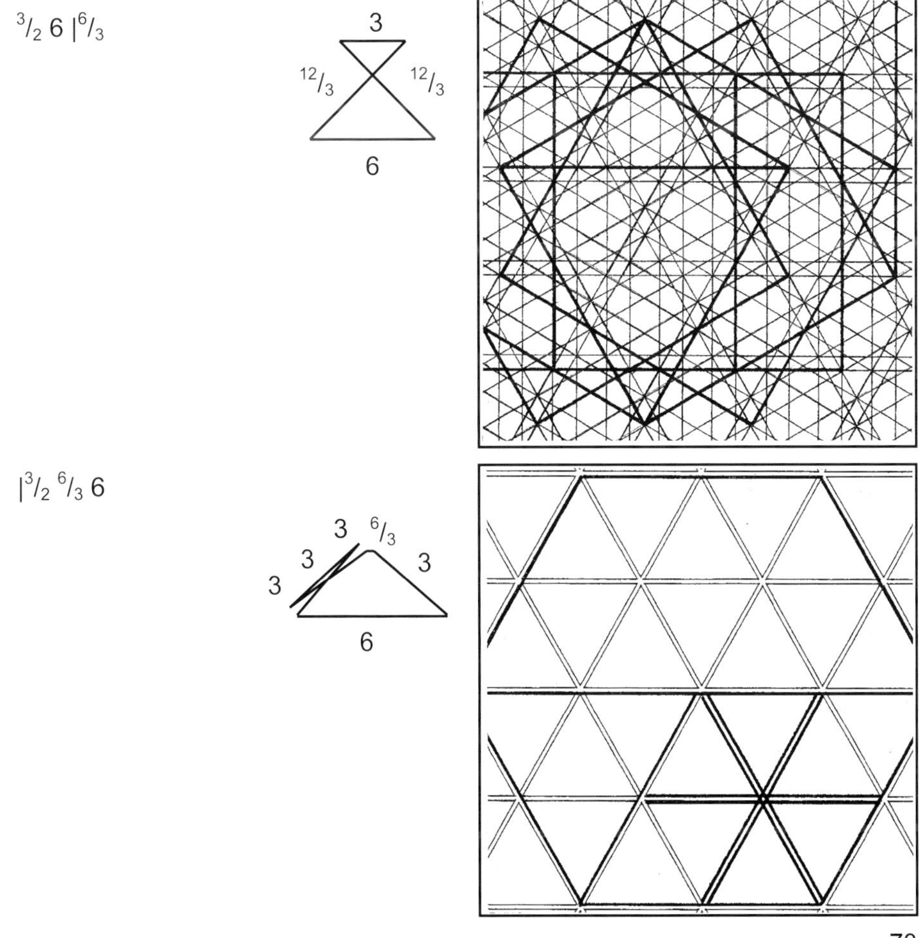

REFERENCES

Coxeter, H.S.M. 1973 *Regular Polytopes* Dover, New York

Coxeter, H.S.M., Longuet-Higgins, M.S. & Miller, J.C.P. 1953 *Uniform Polyhedra* Phil. Trans. R. Soc. Lond. A 246, 401-449

Critchlow, K. 1969 *Order in Space* Thames and Hudson

Grünbaum, B. & Shephard, G.C. 1989 *Tilings and Patterns; An Introduction* W.H.Freeman

Open University 1994 *Tilings* M336 Block 1, Unit IB1

Taylor, P. 1997 *The Complete? Polygon* Nattygrafix

Taylor, P. 1998 *Incomplete Tilings* Nattygrafix

DITRIGONAL VERTEX FIGURES

In the standard classification of the polyhedra, as presented by Coxeter et al., the truncation and quasitruncation of quasi-regular polyhedra excludes certain forms on the grounds that the truncation of polygons {n/d} is not possible when d is even. Such a polygon, say {n/2k}, can in fact be truncated to give a double polygon, for which we have chosen the alternative notation of ²{n/k}, rather than the {2n/2k} that might be expected in the currently accepted notation.

Applying this rule to the truncation and quasitruncation of the quasi-regular polyhedra, we can obtain theoretical triangular vertex figures, which include these double polygon faces, and allow us to construct the forbidden cases of even faced polyhedra. This is done by treating the double polygons as single ones occurring twice, each of which has the other two faces of the triangular vertex figure attached, so that the total vertex figure generally contains six faces and takes a double triangle or ditrigonal form.

The polyhedra thus produced are essentially 'uniform', with regular polygon faces, all arranged equally around identical vertices. They fill in the missing spaces in the standard classification of the even faced polyhedra and also provide the full range of possibilities within the shells of the various rhombihedra and quasirhombihedra that generally comprise the second truncations of the quasi-regular polyhedra.

Most of the chapter that follows was originally presented in 2000 as *Uniform Polyhedra with Ditrigonal Vertex Figures*, a paper read at the Structural Morphology Colloquium that year at Delft. This in turn built on earlier work published in 1995 as Paper Two of *Additions to the Uniform Polyhedra* and that material is now re-presented here with some necessary amendments and additions. The abstract to that paper read:

The truncation and quasitruncation of the regular solids has long been understood and is adequately described by Coxeter, Longuet-Higgins and Miller (1953). The truncation and quasitruncation of the quasi-regular solids, however, remains a confused area limited by their restriction on uniform polyhedra of only two faces being permitted per edge. Skilling (1974) enlarged this definition to include an even number of faces per edge, which will permit the inclusion of several new truncated forms as uniform polyhedra if their vertex figures are correctly interpreted. These are now examined.

TRUNCATING POLYHEDRA

Even faced polyhedra are generally those that are given the Wythoff symbol x y z|, and can be considered as the truncations of the quasi-regular polyhedra x| y z, y| x z or z| x y. If one of the latter two symbols in the quasi-regular is a {2}, the polyhedron being truncated is actually a regular polyhedron, for which we have to truncate the edges as well to produce the {4} in the even faced polyhedron.

For any symmetry family (x y z), we can obtain not only the even faced polyhedron x y z|, but three more such derived from the colunar triangles (x' y' z), (x y' z') and (x' y z'), namely x' y' z|, x y' z'| and x' y z'|. The general even faced polyhedron x y z|, will have a triangular vertex figure consisting of three different faces, t{x}, t{y} and t{z}.

The even faced polyhedra are so called because the truncated polygons forming their faces always have an even number of edges. This property also holds for our double polygons, however Coxeter et al. rejected such polygons resulting from truncating even denominator polygons. They thus discarded a number of possible 'uniform' even faced polyhedra, those illustrated in what follows.

DOUBLE POLYGON VERTICES

When one of the truncated polygons of a theoretical even faced polyhedron is double, all is not lost. By simply treating the double polygon as being a single one there twice, we can construct a six edged vertex figure consisting of two triangles sharing a base. There are fifteen of these, which we will denote as type 2, as against the eight type 1 single triangle versions that Coxeter et al. accepted. A type 2 vertex is effectively visited twice in making up the polyhedron, but generally there are half as many of these vertices compared to the related polyhedra and each vertex has two single edges and two double edges meeting there.

When two of the truncated polygons of a theoretical even faced polyhedron are double (i.e. derived from truncating two even denominator polygons), we can again treat them as singles there twice and obtain multiple vertex figures which take one of two further forms. These were all included by Coxeter et al. in their Table 6 listing 'Degenerate cases of Wythoff's construction'.

Three cases yield a sixfold vertex figure that fits within the shell of the Small Ditrigonal Icosidodecahedron, denoted as type 3, whilst in five special cases the angle between the two double polygon edges of the vertex figure is 90° and we get an inscribed version of a quasi-regular polyhedron with a fourfold vertex figure, type 4. Types 3 and 4 both have triple edges, six and four respectively at each vertex. Typical examples of all four vertex types are shown opposite, marked to show the multiplicity of edges.

76

The table overleaf describes the properties of all thirty one possible even faced polyhedra, generally with three different types of face. The eight of type 1 have been more than adequately described by Coxeter et al. and are illustrated to good effect by Wenninger's models. The five of type 4 are simply multiple quasi-regular polyhedra with their inscribed internal faces present within, as is one of type 3 with some of its internal faces within, the Quasiquasitruncated Icosidodecahedron, $^3/_2\ 2\ ^5/_4|$. The remaining seventeen of types 2 and 3 however, are different from any of the seventy five polyhedra currently acknowledged as uniform, and it seems have only previously been described in the present author's earlier publications.

Not strictly compounds, these polyhedra do have multiple vertices, which are all identical within each polyhedron and make use of only regular polygon faces, albeit some of them double ones where even denominator polygons have been truncated. Whilst their edges are variously single, double or treble, depending on whether two, four or six faces share them, it is always an even number. They are therefore comparable to the Great Disnub Dirhombidodecahedron, with single and double edges, that Skilling proposed adding to the list as the seventy sixth uniform polyhedron, at the end of his paper confirming Coxeter's standard enumeration of the uniform polyhedra.

RHOMBIHEDRAL CONNECTIONS

Each of the twenty three polyhedra presented here can also be seen to fit within the shell of one of the rhombihedra, but they are not the same as the other accepted polyhedra that do this. In general every possible rhombihedral shell can now be seen to produce two such even faced polyhedra, by missing out one of the two parallel edges in the vertex figure and treating the other as a double polygon edge. Indeed those of type 3 are particularly interesting as they are essentially double polygon versions of three rhombihedra also missing from the standard classification: $3\ ^5/_2\ |2$ (Great Rhombicosidodecahedron), $^3/_2\ 5\ |2$ (Quasirhombicosidodecahedron) and $5\ ^5/_3\ |2$ (Quasirhombidodecadodecahedron).

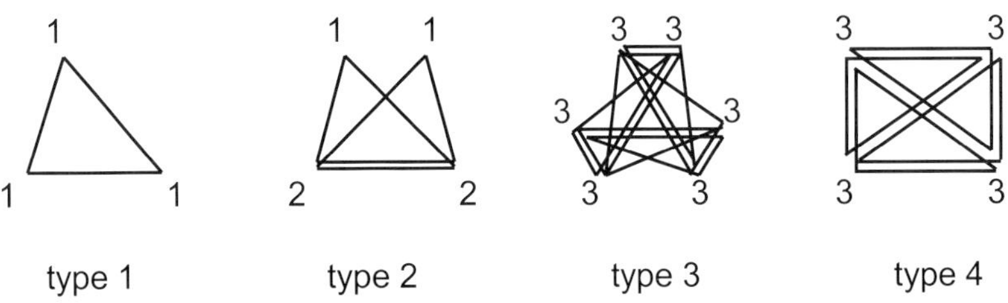

Typical vertex figures showing multiplicity of edges

symbol	type	vertices	edges	faces	fits shell of:	C (W)
$3\ 2\ 3\vert$	1	24	36	$4\{6\}\ 6\{4\}\ 4\{6\}$		20 (7)
$^3/_2\ 2\ 3\vert$	2	12x2	12+(12x2)	$4\{6\}\ 6\{4\}\ 4^2\{3\}$	$3\ 3\ \vert2$	
$^3/_2\ 2\ ^3/_2\vert$	4	6x4	12x3	$4^2\{3\}\ 6\{4\}\ 4^2\{3\}$	$3\ 3\ \vert^3/_2$	
$3\ 2\ 4\vert$	1	48	72	$8\{6\}\ 12\{4\}\ 6\{8\}$		23 (15)
$3\ 2\ ^4/_3\vert$	1	48	72	$8\{6\}\ 12\{4\}\ 6\{^8/_3\}$		67 (93)
$^3/_2\ 2\ 4\vert$	2	24x2	24+(24x2)	$8^2\{3\}\ 12\{4\}\ 6\{8\}$	$3\ 4\ \vert2$	
$^3/_2\ 2\ ^4/_3\vert$	2	24x2	24+(24x2)	$8^2\{3\}\ 12\{4\}\ 6\{^8/_3\}$	$3\ 4\ \vert^4/_3$	
$3\ 2\ 5\vert$	1	120	180	$20\{6\}\ 30\{4\}\ 12\{10\}$		31 (16)
$3\ 2\ ^5/_4\vert$	2	60x2	60+(60x2)	$20\{6\}\ 30\{4\}\ 12^2\{^5/_2\}$	$5\ ^5/_2\ \vert2$	
$^3/_2\ 2\ 5\vert$	2	60x2	60+(60x2)	$20^2\{3\}\ 30\{4\}\ 12\{10\}$	$3\ 5\ \vert2$	
$^3/_2\ 2\ ^5/_4\vert$	3	20x6	60x3	$20^2\{3\}\ 30\{4\}\ 12^2\{^5/_2\}$	$3\ ^5/_2\ \vert2$	
$3\ 2\ ^5/_2\vert$	2	60x2	60+(60x2)	$20\{6\}\ 30\{4\}\ 12^2\{5\}$	$5\ ^5/_2\ \vert2$	
$3\ 2\ ^5/_3\vert$	1	120	180	$20\{6\}\ 30\{4\}\ 12\{^{10}/_3\}$		87 (108)
$^3/_2\ 2\ ^5/_2\vert$	3	20x6	60x3	$20^2\{3\}\ 30\{4\}\ 12^2\{5\}$	$^3/_2\ 5\ \vert2$	
$^3/_2\ 2\ ^5/_3\vert$	2	60x2	60+(60x2)	$20^2\{3\}\ 30\{4\}\ 12\{^{10}/_3\}$	$3\ ^5/_2\ \vert^5/_3$	
$5\ 2\ ^5/_2\vert$	2	60x2	60+(60x2)	$12\{10\}\ 30\{4\}\ 12^2\{5\}$	$3\ 5\ \vert2$	
$5\ 2\ ^5/_3\vert$	1	120	180	$12\{10\}\ 30\{4\}\ 12\{^{10}/_3\}$		75 (98)
$^5/_4\ 2\ ^5/_2\vert$	3	20x6	60x3	$12^2\{^5/_2\}\ 30\{4\}\ 12^2\{5\}$	$5\ ^5/_3\ \vert2$	
$^5/_4\ 2\ ^5/_3\vert$	2	60x2	60+(60x2)	$12^2\{^5/_2\}\ 30\{4\}\ 12\{^{10}/_3\}$	$3\ ^5/_2\ \vert^5/_3$	
$4\ 2\ ^4/_2\vert$	2	24x2	24+(24x2)	$6\{8\}\ 12\{4\}\ 6^2\{4\}$	$3\ 4\ \vert2$	
$^4/_3\ 2\ ^4/_2\vert$	2	24x2	24+(24x2)	$6\{^8/_3\}\ 12\{4\}\ 6^2\{4\}$	$3\ 4\ \vert^4/_3$	
$3\ ^4/_2\ ^3/_2\vert$	4	12x4	24x3	$8\{6\}\ 6^2\{4\}\ 8^2\{3\}$	$3\ 3\ \vert^4/_2$	
$4\ 3\ ^4/_3\vert$	1	48	72	$6\{8\}\ 8\{6\}\ 6\{^8/_3\}$		52 (79)
$3\ ^5/_3\ ^3/_2\vert$	2	60x2	60+(60x2)	$20\{6\}\ 12\{^{10}/_3\}\ 20^2\{3\}$	$3\ 5\ \vert^5/_3$	
$5\ 3\ ^5/_4\vert$	2	60x2	60+(60x2)	$12\{10\}\ 20\{6\}\ 12^2\{^5/_2\}$	$3\ ^5/_2\ \vert3$	
$3\ 5\ ^3/_2\vert$	2	60x2	60+(60x2)	$20\{6\}\ 12\{10\}\ 20^2\{3\}$	$3\ ^5/_2\ \vert3$	
$^5/_2\ 3\ ^5/_3\vert$	2	60x2	60+(60x2)	$12^2\{5\}\ 20\{6\}\ 12\{^{10}/_3\}$	$3\ 5\ \vert^5/_3$	
$3\ ^5/_3\ 5\vert$	1	120	180	$20\{6\}\ 12\{^{10}/_3\}\ 12\{10\}$		57 (84)
$3\ ^5/_2\ ^5/_4\vert$	4	30x4	60x3	$20\{6\}\ 12^2\{5\}\ 12^2\{^5/_2\}$	$5\ 5\ \vert^5/_4$	
$^3/_2\ ^5/_2\ 5\vert$	4	30x4	60x3	$20^2\{3\}\ 12^2\{5\}\ 12\{10\}$	$3\ 3\ \vert^5/_2$	
$^3/_2\ ^5/_3\ ^5/_4\vert$	4	30x4	60x3	$20^2\{3\}\ 12\{^{10}/_3\}\ 12^2\{^5/_2\}$	$3\ 3\ \vert^5/_4$	

$^3/_2$ 2 3| $tt^2\{3',3\}$

Additions 2.4a
Kasparian 3

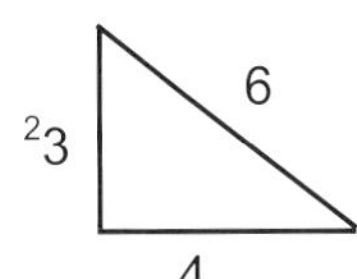

type 2

Triquasitruncated Tetratetrahedron

Fits within shell of 3 3 |2

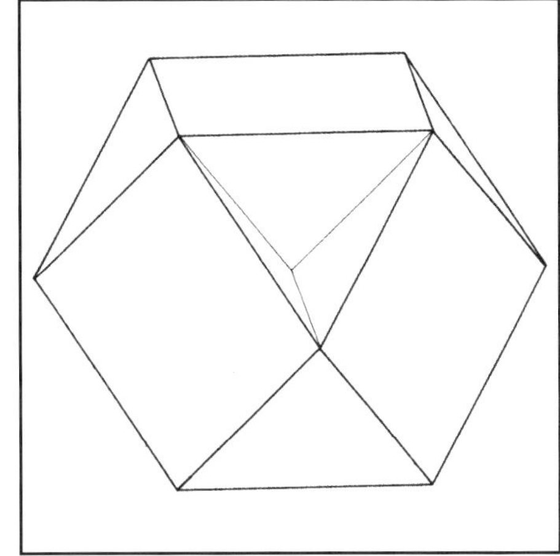

$^3/_2$ 2 4| $tt^2\{3',4\}$

Additions 2.4b
Kasparian 9

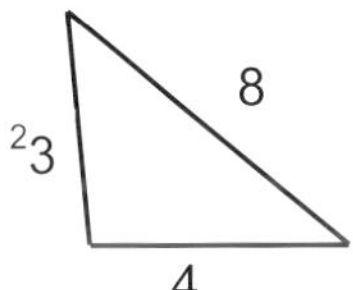

type 2

Triquasitruncated Cuboctahedron

Fits within shell of 3 4 |2

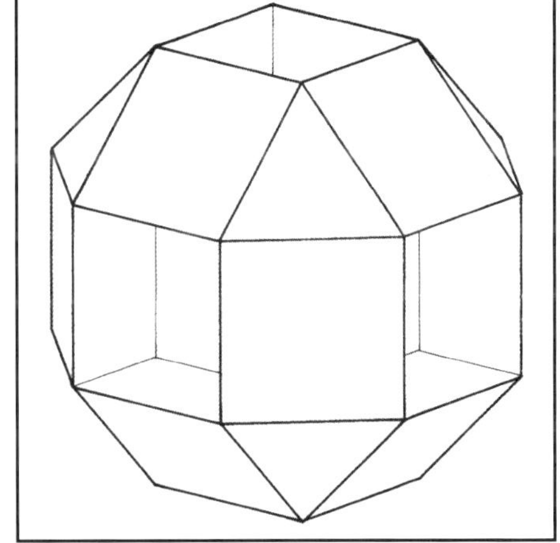

$^3/_2$ 2 $^4/_3$| $tt^2\{3',4'\}$

Additions 2.4c
Kasparian 15

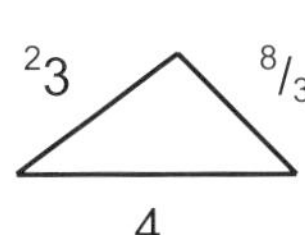

type 2

Quasiquasitruncated Cuboctahedron

Fits within shell of 3 4 |$^4/_3$

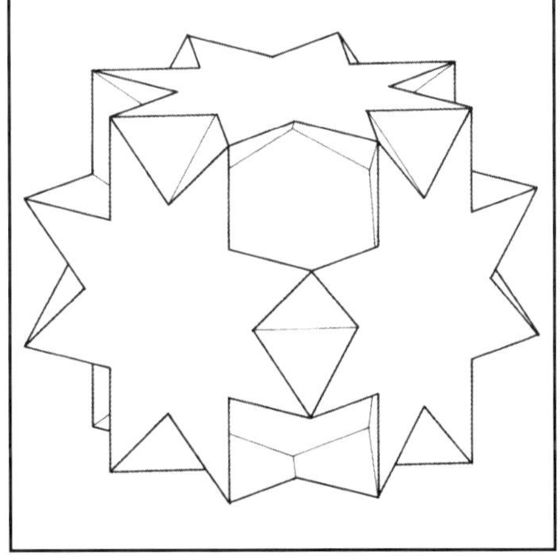

THE (3 2 5) FAMILY

$3\ 2\ ^5/_4|$ $tt^2\{3,5'\}$

Additions 2.4d
Kasparian 18

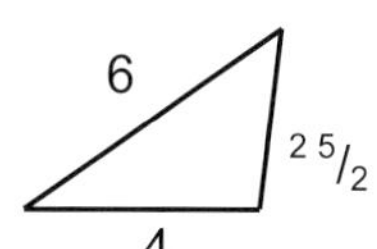

type 2

Pentaquasitruncated
Icosidodecahedron

Fits within shell of $5\ ^5/_2\ |2$

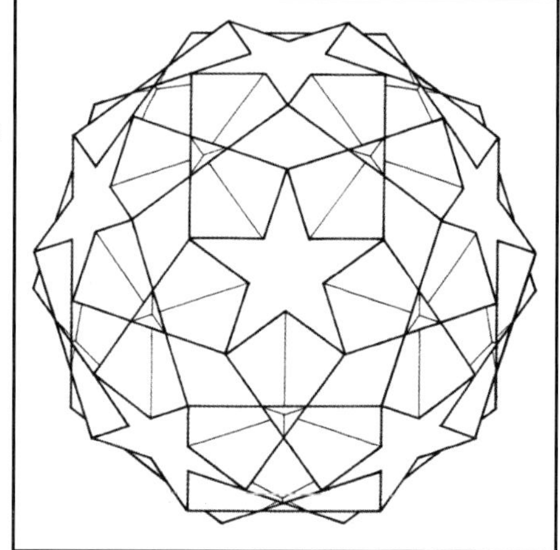

$^3/_2\ 2\ 5|$ $tt^2\{3',5\}$

Additions 2.4e
Kasparian 10

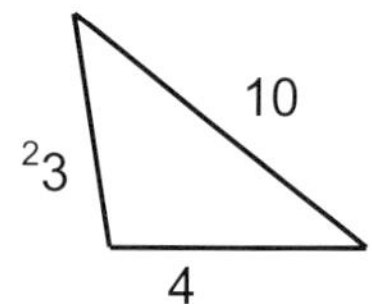

type 2

Triquasitruncated Icosidodecahedron

Fits within shell of $3\ 5\ |2$

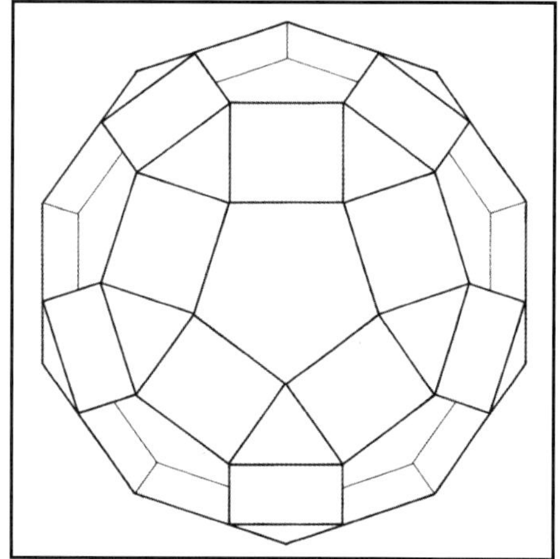

$^3/_2\ 2\ ^5/_4|$ $tt^2\{3',5'\}$

Additions 2.4f
Kasparian 29

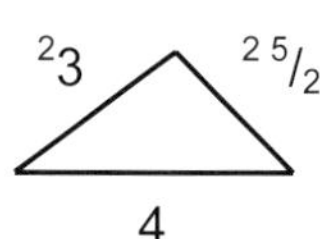

type 3

Quasiquasitruncated
Icosidodecahedron

Fits within shell of $3\ ^5/_2\ |2$
(Great Rhombicosidodecahedron)

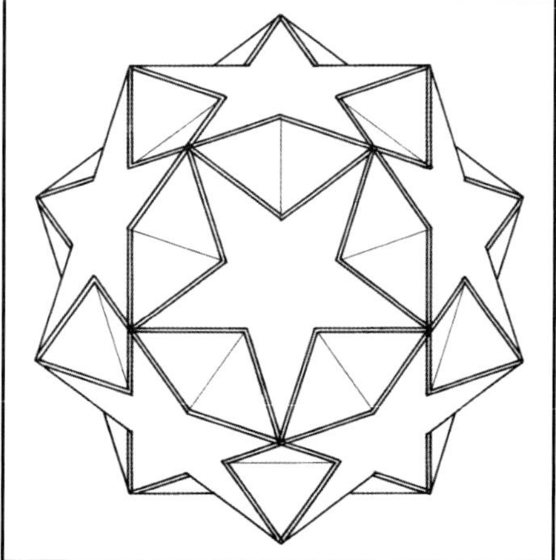

$3\ 2\ ^5/_2|$ $tt^2\{3,^5/_2\}$

Additions 2.4g
Kasparian 7

type 2

Truncated Great Icosidodecahedron

Fits within shell of $5\ ^5/_2\ |2$

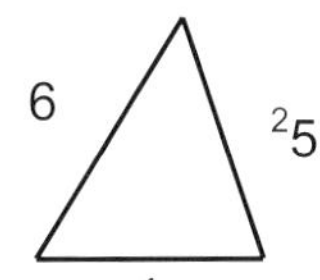
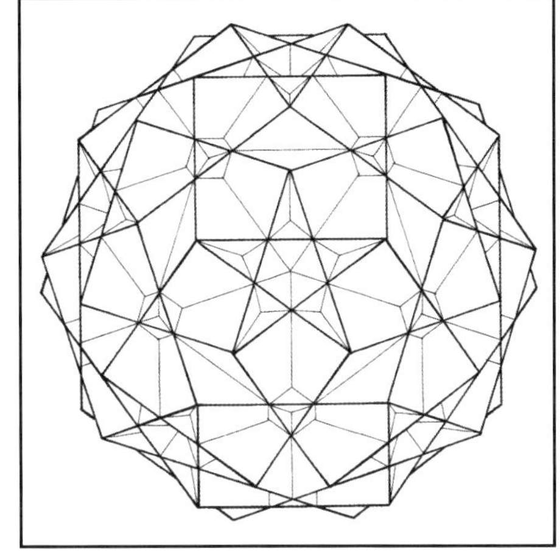

$^3/_2\ 2\ ^5/_2|$ $tt^2\{3',^5/_2\}$

Additions 2.4h
Kasparian 2

type 3

Triquasitruncated Great
Icosidodecahedron

Fits within shell of $^3/_2\ 5\ |2$
(Quasirhombicosidodecahedron)

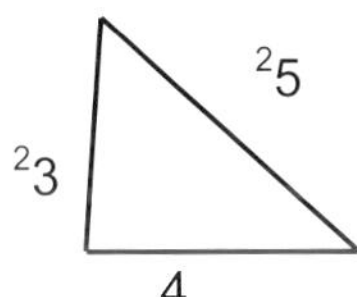
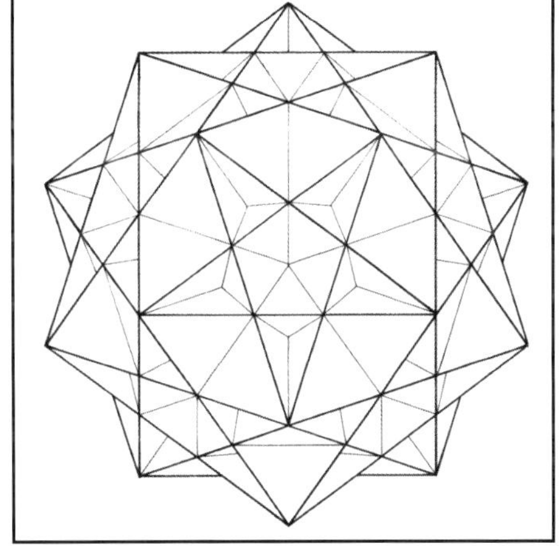

$^3/_2\ 2\ ^5/_3|$ $tt^2\{3',^5/_2{}'\}$

Additions 2.4i
Kasparian 12

type 2

Quasiquasitruncated Great
Icosidodecahedron

Fits within shell of $3\ ^5/_2\ |^5/_3$

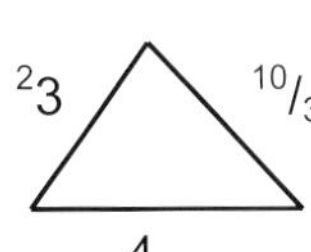
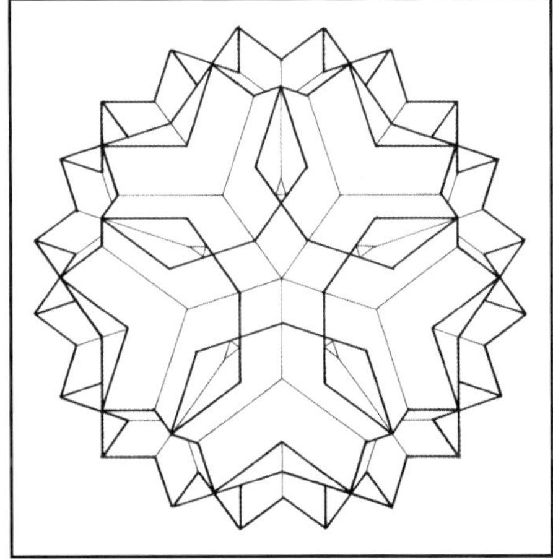

THE (5 2 $^5/_2$) FAMILY

5 2 $^5/_2$| tt^2{5,$^5/_2$}

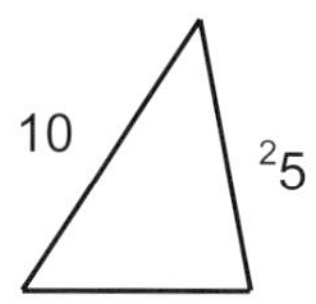

Additions 2.4j
Kasparian 8

type 2

Truncated Dodecadodecahedron
(see p.28)

Fits within shell of 3 5 |2

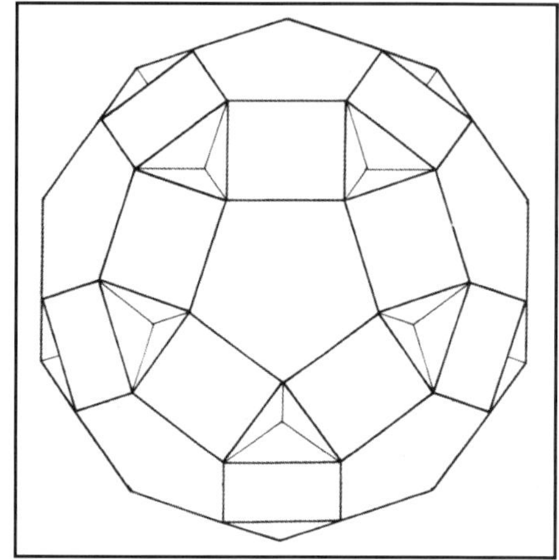

$^5/_4$ 2 $^5/_2$| tt^2{5',$^5/_2$}

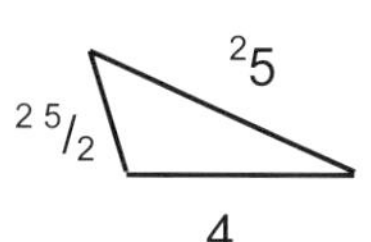

Additions 2.4k
Kasparian 21

type 3

Pentaquasitruncated
Dodecadodecahedron
(see p.28)

Fits within shell of 5 $^5/_3$ |2
(Quasirhombidodecadodecahedron)

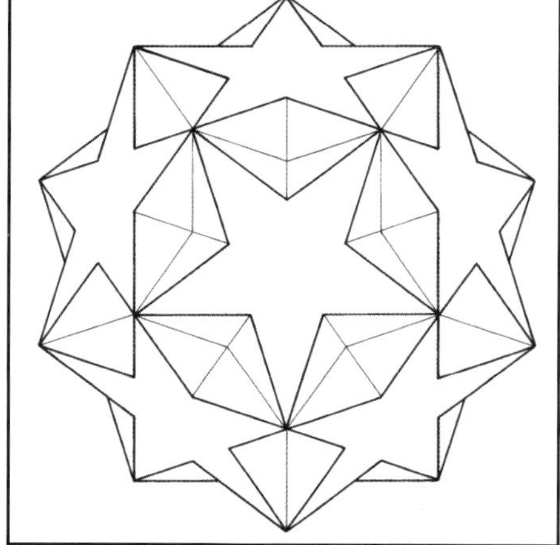

$^5/_4$ 2 $^5/_3$| tt^2{5',$^5/_2$'}

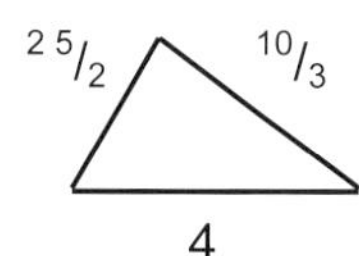

Additions 2.4l
Kasparian 13

type 2

Quasiquasitruncated
Dodecadodecahedron
(see p.29)

Fits within shell of 3 $^5/_2$ |$^5/_3$

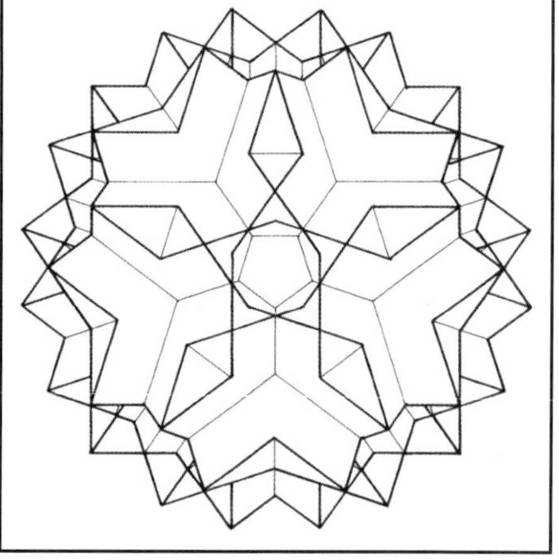

4 2 $^4/_2$| tt^2\{4,$^4/_2$\}

Additions 2.4m
Kasparian -

type 2

Truncated Hexahexahedron
(see p.56)

Fits within shell of 3 4 |2

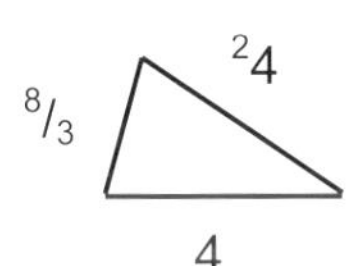

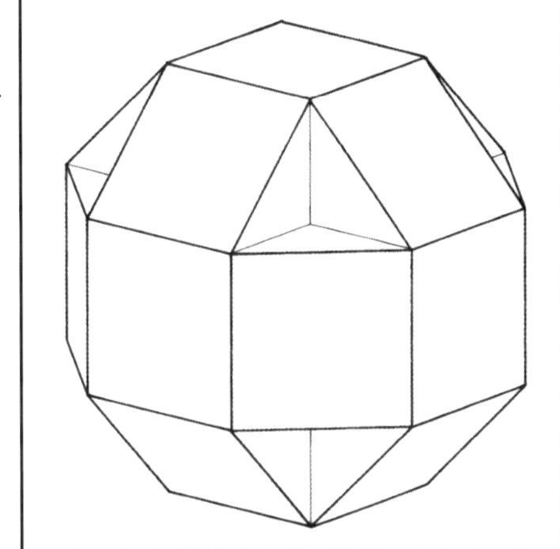

$^4/_3$ 2 $^4/_2$| tt^2\{4',$^4/_2$\}

Additions 2.4n
Kasparian -

type 2

Quasitruncated Hexahexahedron
(see p.56)

Fits within shell of 3 4 |$^4/_3$

There are no further truncations or
quasitruncations in this family since
\{$^4/_2$\} is self dual.

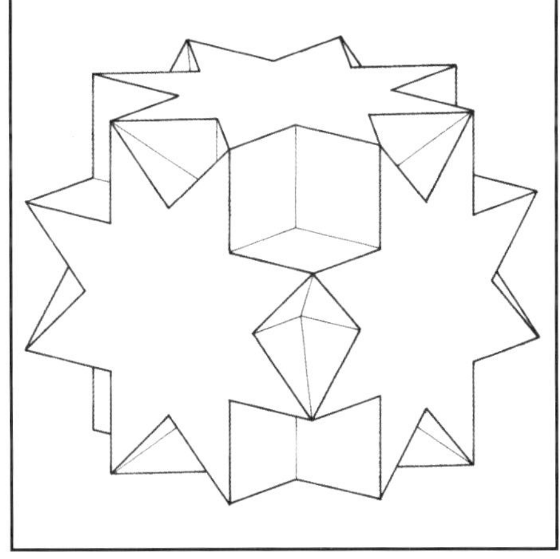

KASPARIAN POLYHEDRA

The nine even faced polyhedra that follow were not originally published as
part of *'Additions to the Uniform Polyhedra'* (1995) as they mostly belong to
sub-families and a super-family.   All were later published in either *'The
Simpler? Polyhedra'* (1999) or *'The Star & Cross Polyhedra'* (2000).

Meanwhile in 1998 Raffi J. Kasparian started work on a paper that covered
pretty much the same ground: his 'Kasparian Polyhedra' include all of the
ditrigonals shown here with the exception of those involving a \{$^4/_2$\} cross
polygon. His work examined the possible vertex configurations that can be
used repeatedly and consistently to generate virtual polyhedral shells using
the programme: www.quantimegroup.com/solutions/Archimedean.

DOUBLE FAMILIES

3 $^4/_2$ $^3/_2$|

Additions -
Kasparian -

type 4

Quasiquasitruncated Stella Octangula

Inscribed Cuboctahedron

Fits within shell of 3 3 |$^4/_2$

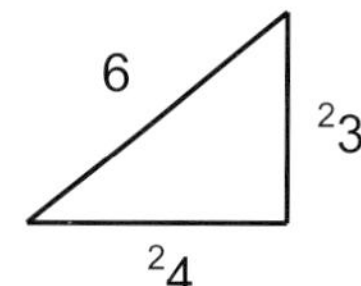
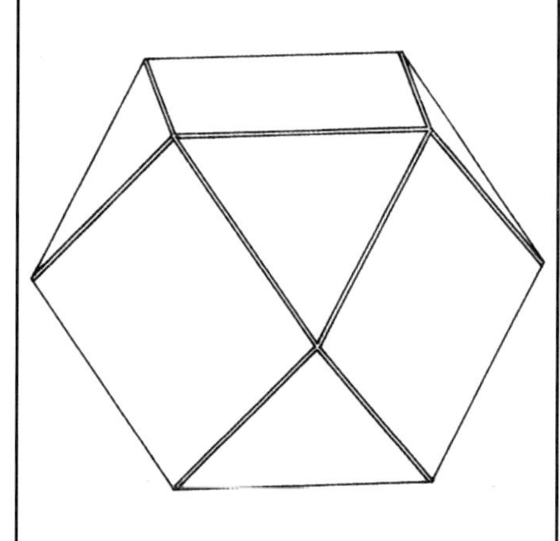

3 $^5/_3$ $^3/_2$|

Additions -
Kasparian 16

type 2

Quasiquasitruncated Small Ditrigonal Icosidodecahedron

Fits within shell of 3 5 |$^5/_3$

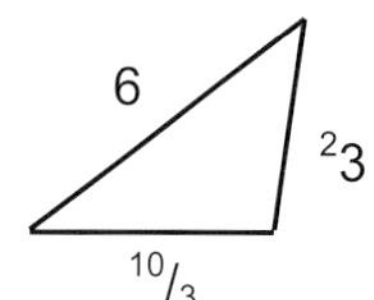
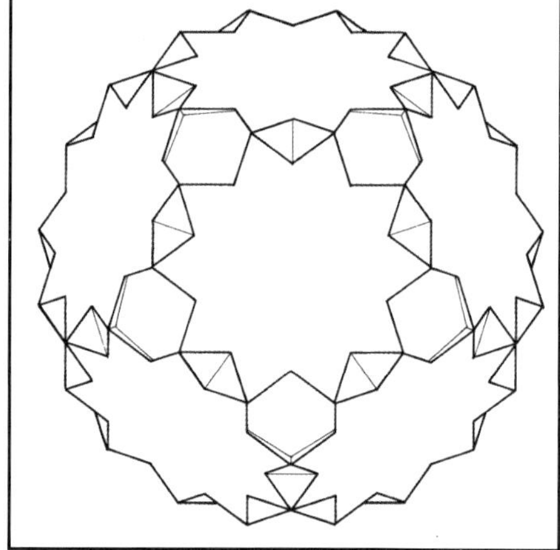

5 3 $^5/_4$|

Additions -
Kasparian 20

type 2

Quasiquasitruncated Inscribed Icosahedron

Fits within shell of 3 $^5/_2$ |3

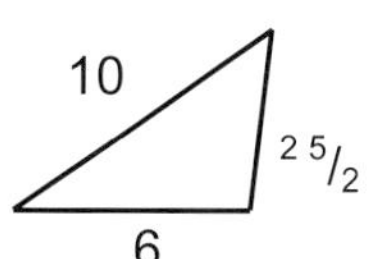
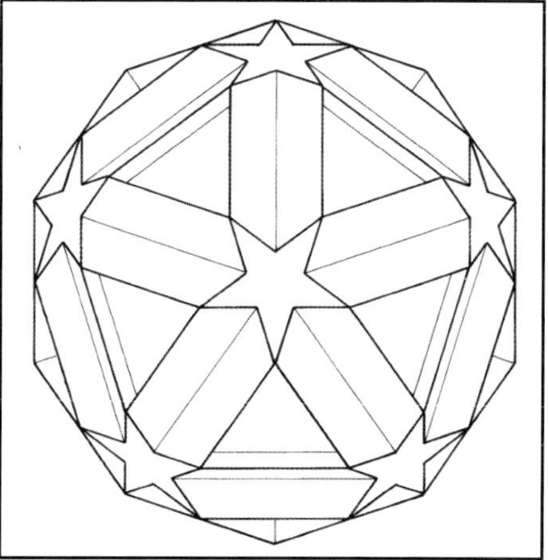

$^3/_2$ 2 $^3/_2|$

Additions -
Kasparian 1

type 4

Quasiquasitruncated
Tetratetrahedron

Inscribed Octahedron

Fits within shell of 3 3 $|^3/_2$

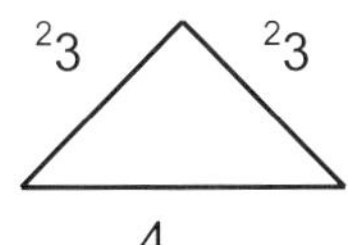

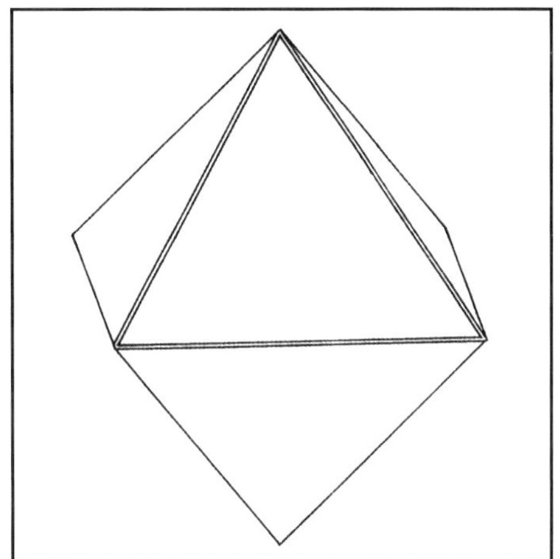

3 5 $^3/_2|$

Additions -
Kasparian 6

type 2

Quasiquasitruncated Great
Ditrigonal Icosidodecahedron

Fits within shell of 3 $^5/_2$ |3

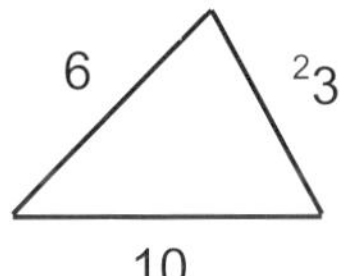

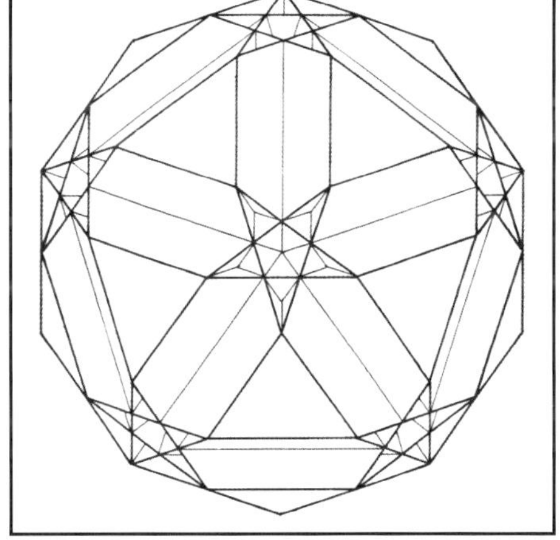

$^5/_2$ 3 $^5/_3|$

Additions -
Kasparian 17

type 2

Quasiquasitruncated Inscribed Small
Stellated Dodecahedron

Fits within shell of 3 5 $|^5/_3$

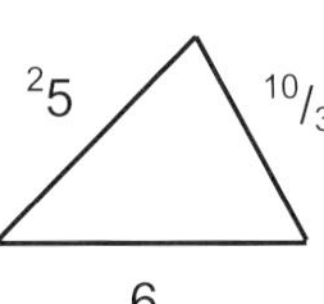

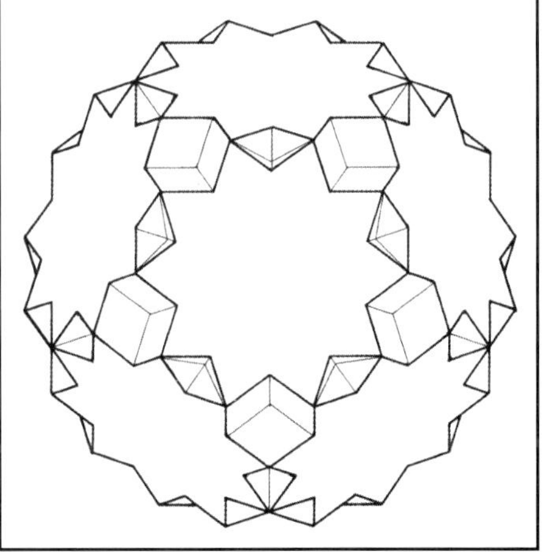

THE $(3\ {}^5/_3\ 5)$ SUPER-FAMILY

$3\ {}^5/_2\ {}^5/_4|$

Additions -
Kasparian 19

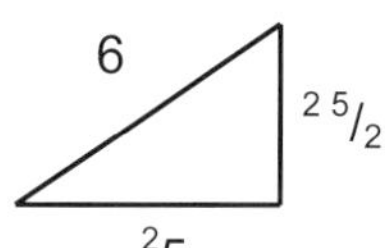

type 4

Inscribed Dodecadodecahedron

Fits within shell of $5\ 5\ |{}^5/_4$

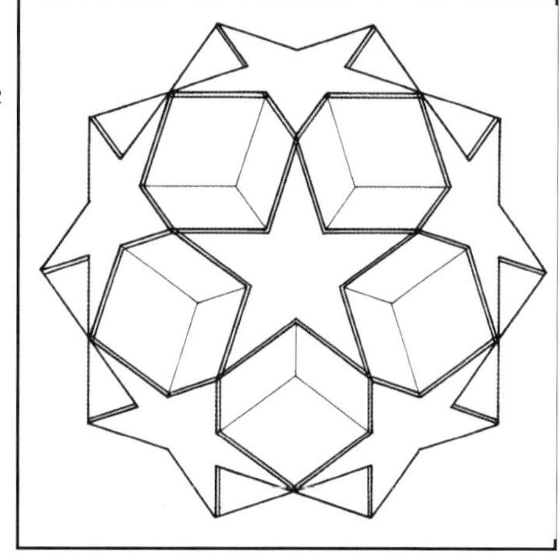

$^3/_2\ {}^5/_2\ 5|$

Additions -
Kasparian 5

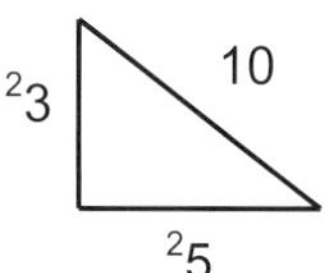

type 4

Inscribed Icosidodecahedron

Fits within shell of $3\ 3\ |{}^5/_2$

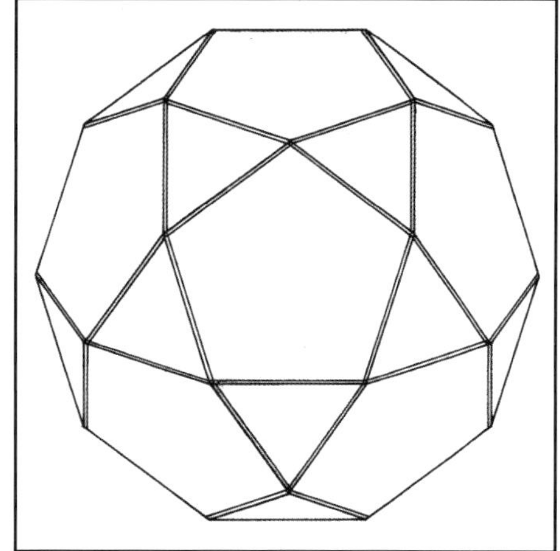

$^3/_2\ {}^5/_3\ {}^5/_4|$

Additions -
Kasparian 14

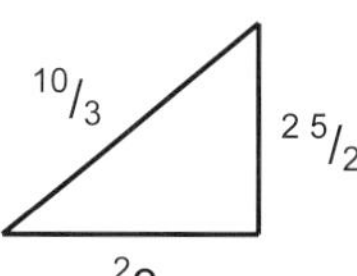

type 4

Inscribed Great Icosidodecahedron

Fits within shell of $3\ 3\ |{}^5/_4$

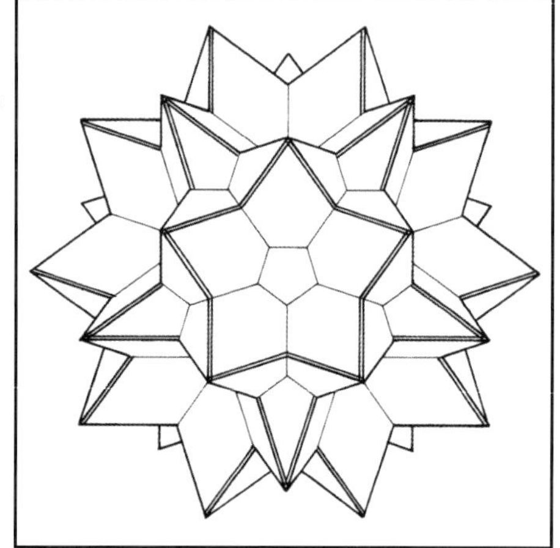

It might be thought that by opening the gates to these new possibilities, there will be a flood of other possible polyhedra seeking consideration as 'uniform'. This is not the case. The truncation of polygons using the alternative notation has shown us only two possibilities, odd or even denominator, the latter producing the double polygons we have used, with no possible extension to trebles, quadruples etc. as might be feared.

Indeed, the not otherwise encountered treble, quadruple etc. polygons seem to be rather useless, purely a product of the currently accepted notation for polygons, where they arise when a polygon {n/d} has n and d not co-prime. In the alternative notation used here such polygons are simply compound and as we have seen with the star and cross tilings and the cross polyhedra, far more useful for constructing new forms.

All twenty three of the polyhedra produced using the double polygons can be assigned to one or another of the already existing symmetry families, as defined by the various Schwarz triangles within known spherical kaleidoscopes. In fact all that we have done is allowed the inclusion of double polygons as true truncations, and thus filled in the missing even faced polyhedra that previously left holes in the standard classification. It is to be hoped that the vocabulary of possible forms is now perhaps complete?

It would seem so since we have also seen that every possible rhombihedral shell can be used to generate two of these ditrigonal even faced polyhedra, by missing out one of the two parallel edges in the vertex figure and treating the other as a double polygon edge. This method produces exactly the same forms as our consideration of the missing forms in the standard classification. Indeed Kasparian's method has come up with exactly the same forms too (his nos. 1-3, 5-10, 12-21 and 29), apparently lacking only the three that involve the double square truncation of $\{^4/_2\}$, which is understandable.

REFERENCES

Coxeter, H.S.M. 1973 *Regular Polytopes* Dover, New York

Coxeter, H.S.M., Longuet-Higgins, M.S. & Miller, J.C.P. 1953 *Uniform Polyhedra* Phil. Trans. R. Soc. Lond. A 246, 401-449

Critchlow, K. 1969 *Order in Space* Thames and Hudson

Cundy, H.M. & Rollett, A.P. 1951 *Mathematical Models* Oxford U P

Holden, A. 1971 *Shapes, Space, and Symmetry* Columbia U P

Kasparian, R.J. & Petillo, A.E. 2017 *Introducing the Kasparian Constructions* Bridges 2017 Conference Proceedings

Skilling, J. 1974 *The Complete Set of Uniform Polyhedra* Phil. Trans. R. Soc. Lond. A 278, 111-135

Taylor, P. 1995 *Additions to the Uniform Polyhedra* Nattygrafix

Taylor, P. 1997 *The Complete? Polygon* Nattygrafix

Taylor, P. 1998 *Incomplete Tilings* Nattygrafix

Taylor, P. 1999 *The Simpler? Polyhedra* Nattygrafix

Taylor, P. 2000 *The Star & Cross Polyhedra* Nattygrafix

Wenninger, M.J. 1971 *Polyhedron Models* Cambridge U P

Wenninger, M.J. 1983 *Dual Models* Cambridge U P

We have thus far explored the results of making new Schwarz triangles from varying numbers of the smaller Möbius triangles within the regular tiling and polyhedral symmetry families. We found that the compound polygons $\{^6/_2\}$, $\{^4/_2\}$ and $\{^6/_3\}$, as described by the alternative notation used here, do indeed allow us to flesh out the members of various new symmetry families of cross and star tilings and cross polyhedra, all analogous to existing families of tilings or polyhedra, as shown in the preceding chapters.

We have also been able to interpret the double polygons, which result from truncating even denominator polygons in the alternative notation, as just that - polygons present twice in a ditrigonal vertex figure.

We now need to step up a level and see if these same new cross or star based forms can produce larger assemblies equivalent to the closepacks or hypersolids that the polyhedra produce when assembled together. Not surprisingly we find that there are forms using these polygons that follow the same rules of truncation between dual pairs as do the closepacks and hypersolids. We will call these crosspacks and cross hypersolids respectively, this chapter will deal with the former, the next the latter.

We will start with a little recap taking us up through the various dimensional levels. Certain regular polytopes are found to repeat with analogues at any number of dimensions. The most simple are accordingly called the 'simplex polytopes', $\alpha_n = \{3^{n-1}\}$, which form a series comprising triangle α_2 $\{3\}$, tetrahedron α_3 $\{3,3\}$, hypertetrahedron α_4 $\{3,3,3\}$ etc.

Another series are dubbed 'measure polytopes' $\gamma_n = \{4,3^{n-2}\}$, which form a series comprising square γ_2 $\{4\}$, cube γ_3 $\{4,3\}$, hypercube γ_4 $\{4,3,3\}$ etc. These provide our basic units of measure in any number of dimensions such as area, volume etc.

A third series are the duals of the measure polytopes, which we will sensibly call the 'dual polytopes' $\beta_n = \{3^{n-2},4\}$, which form a series comprising square β_2 $\{4\}$, octahedron β_3 $\{3,4\}$, octahypertetrahedron β_4 $\{3,3,4\}$ etc.

In the accepted version of polyhedral geometry, where the cross polygon $\{^4/_2\}$ is deemed not to exist, these 'dual polytopes' have been misnamed as the 'cross polytopes', a term we will need to reserve for describing the polygons, polyhedra and hypersolids etc. that do use the $\{^4/_2\}$ polygon face and form the bulk of the content of this volume.

A further series dubbed 'honeycombs' $\delta_n = \{4,3^{n-3},4\}$ fill space at different dimensions and form a series comprising square tiling δ_3 $\{4,4\}$, cubic closepack δ_4 $\{4,3,4\}$, hypercube honeycomb δ_5 $\{4,3,3,4\}$ etc.

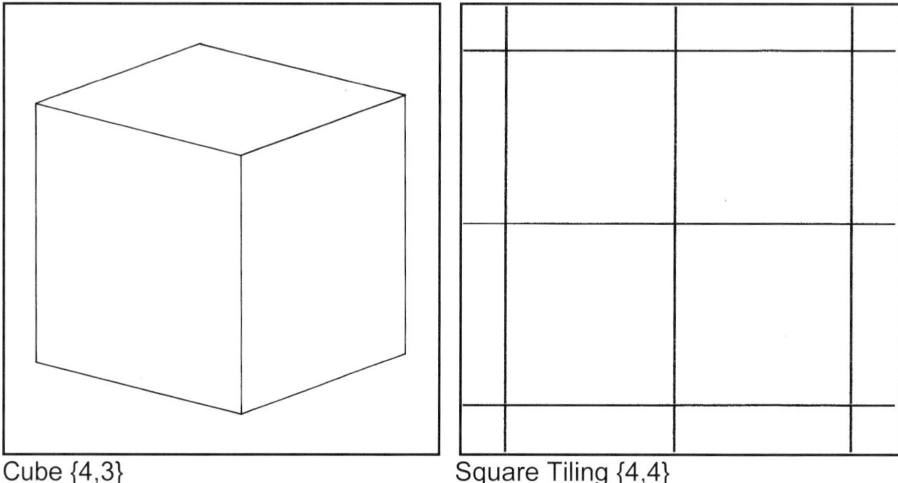

Cube {4,3}

Square Tiling {4,4}

We tend to measure the world by the square γ_2 {4} and volume by the cube γ_3 {4,3}, this last comprising squares arranged three per vertex. With four squares per vertex, rather than wrapping around a volume we fill the plane with an infinite square tiling δ_3 {4,4}. This is self-dual since another square tiling can be found by joining the centres of the original square faces.

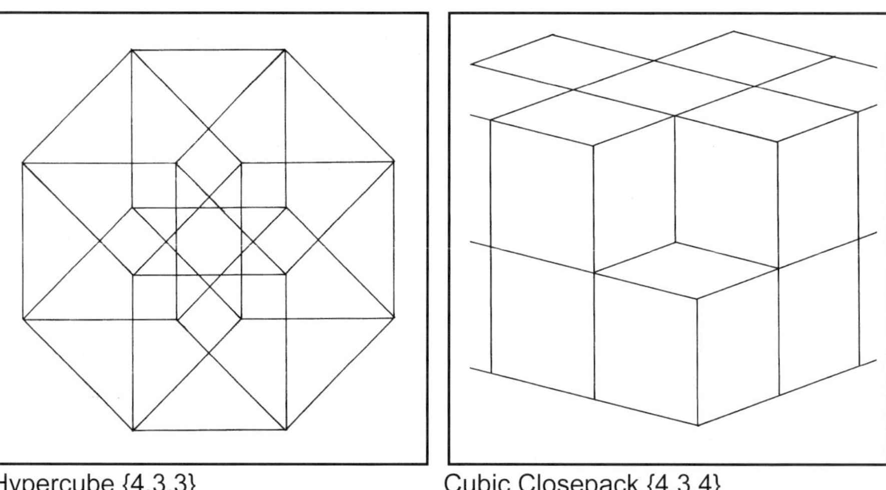

Hypercube {4,3,3}

Cubic Closepack {4,3,4}

Similarly up a level of organisation, cubes arranged three per edge give us a hypercube γ_4 {4,3,3} that wraps up a section of hyperspace, whilst with four cubes per edge it yields a cubic closepack δ_4 {4,3,4} that fills three dimensions to infinity. This is also self-dual as another cubic closepack can be found by joining together the centres of the original cubes.

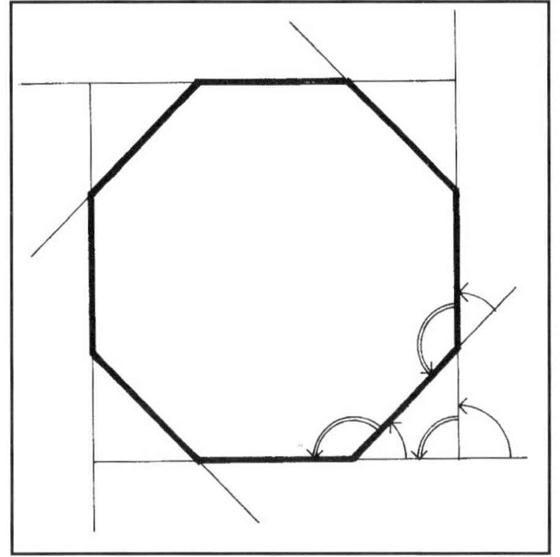

Truncated Square t{4}

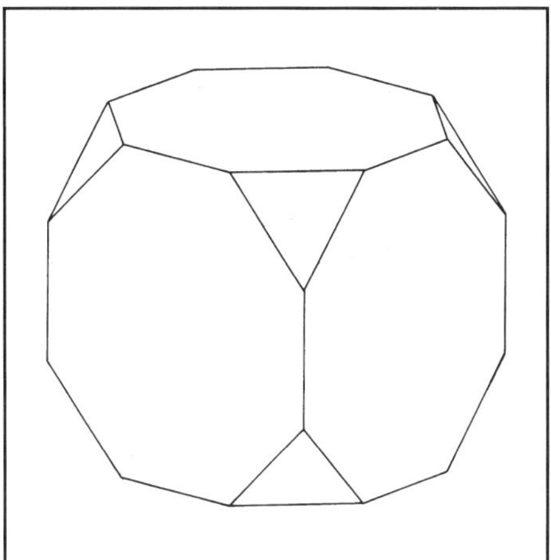

Truncated Cube t{4,3}

Within this simple world we can truncate by growing new edges, faces or solids around a vertex leading to semi-regular forms with varying faces. A square {4} truncates to an octagon {8} with half the edges original, half new grown. Similarly a cube truncates to a truncated cube with six octagonal faces (original) and eight triangular faces (new).

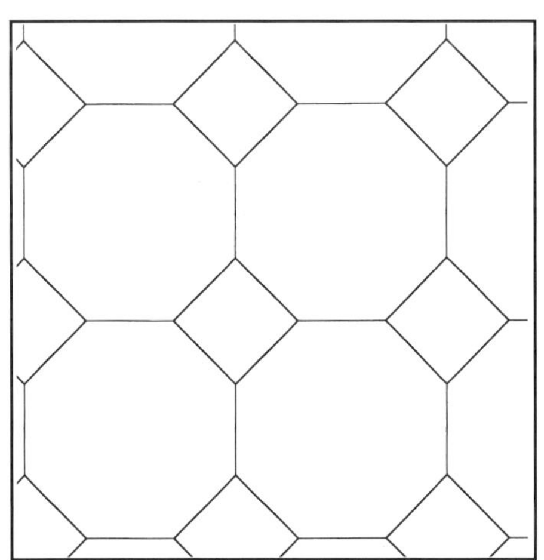

Truncated Square Tiling t{4,4}

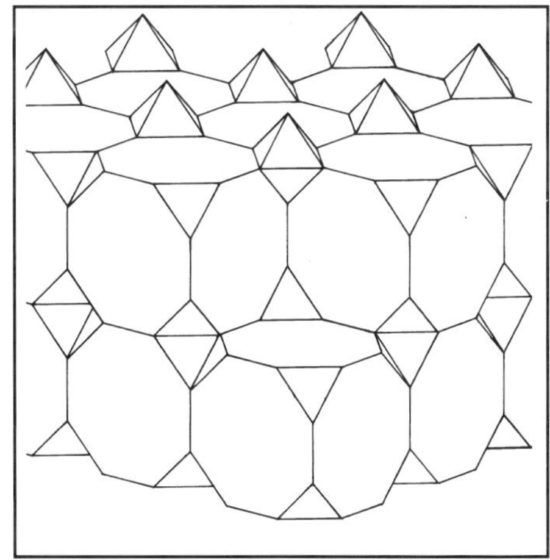

Truncated Cubic Closepack t{4,3,4}

Similarly truncating a square tiling produces a tiling of octagons (original) and squares (new), whilst truncating a cubic closepack yields a closepack comprising truncated cubes (original) and octahedra (new). Thus far this is fairly simple polytope geometry.

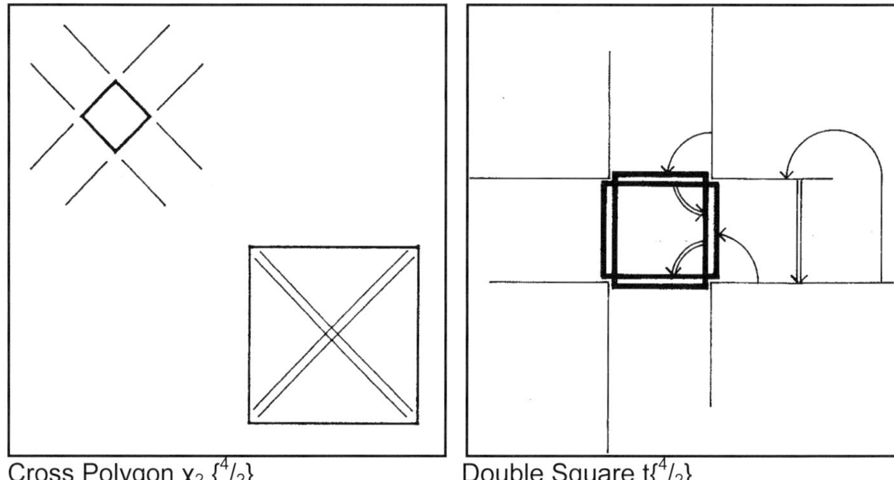

Cross Polygon χ₂ {⁴/₂}   Double Square t{⁴/₂}

The cross polygon $\{^4/_2\}$ can be obtained by stellating a square (extending its faces until they meet) or by inscribing a new polygon within. It has no area, but like the square it has four vertices and four edges and an internal angle (0°) given by the formula for a general polygon $\{n\}$, $\theta = 180 - (360/n)°$. Its truncation has eight vertices and eight edges (four original, four new), arranged as a double square $^2\{4\}$.

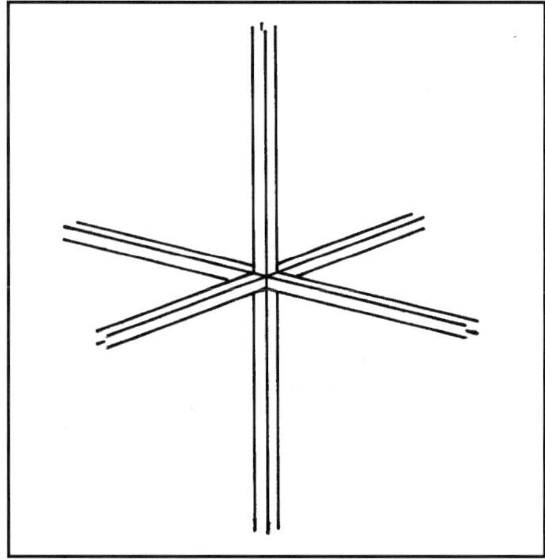

Stellated Hexahedron χ₃ {⁴/₂,4}

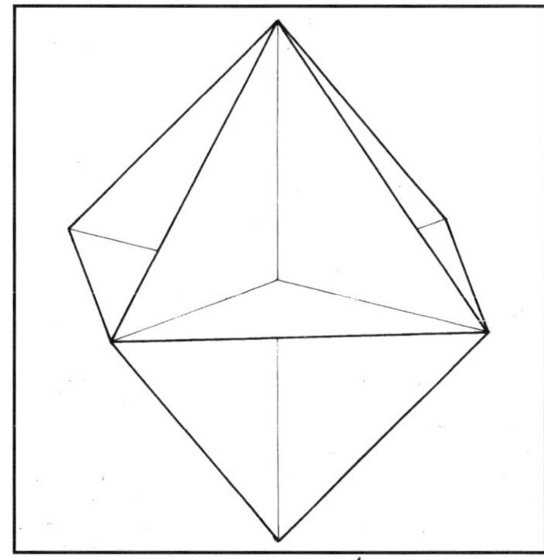

Great Hexahedron φ₃ {4,⁴/₂}

As we have already seen, the cross polygon $\{^4/_2\}$ can be arranged to form a cross polyhedron, the stellated hexahedron $\{^4/_2,4\}$, by placing four of them at each vertex. It has a dual, the great hexahedron $\{4,^4/_2\}$, with four squares arranged around each $^4/_2$ shaped vertex, an inscription of the octahedron. Both of these comprise six faces and have six vertices and twelve edges.

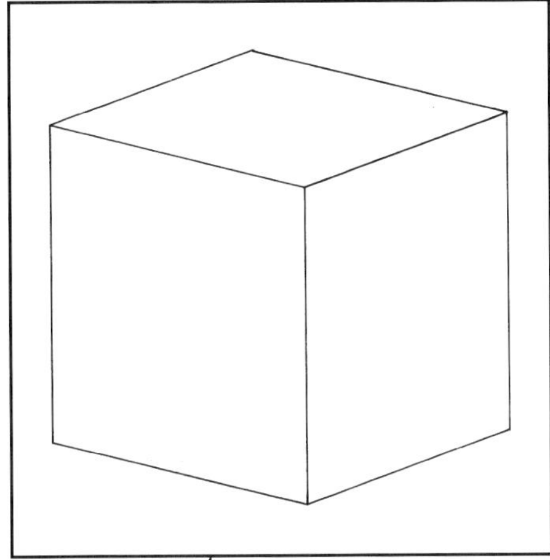

Triple Cube t{$^4/_2$,4}

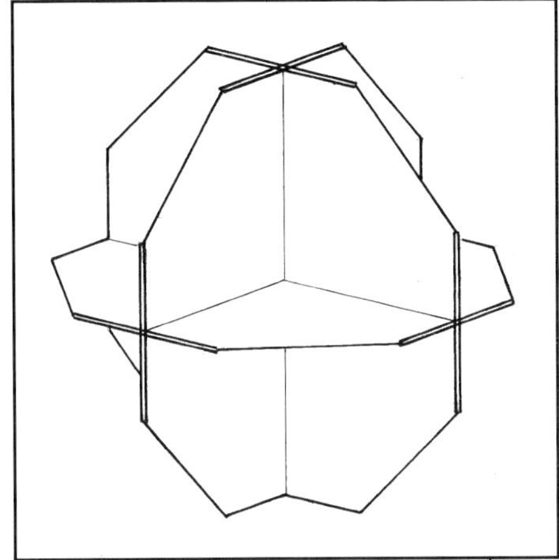

Truncated Great Hexahedron t{4,$^4/_2$}

These cross polyhedra can be truncated in the usual way to provide a series of intermediate forms. The stellated hexahedron truncates to produce three coincident cubes, analogous to the double square produced a dimension lower, whilst the great hexahedron truncates to a form comprising six octagons in back to back pairs, linked together by new cross polygon faces.

Going one dimension higher again, we can envisage a 'tetratile crosspack' ζ_4 {4,4,$^4/_2$} comprising square tilings arranged $^4/_2$ per edge. This is essentially a cubic closepack without the solids present, just the planes filled with squares separating them, and like the cubic closepack it extends to fill all of a three dimensional world. The Schläfli notation used here has been extended so that {x,y,z} represents both polyhedra or tilings {x,y} arranged z per edge and also faces {x} arranged around a polyhedral vertex figure {y,z}.

The tetratile crosspack's dual ε_4 {$^4/_2$,4,4} is a little different comprising an infinite number of stellated hexahedra, four per edge, sharing just six vertices. Rather than a polyhedron, its vertex figure is an infinitely extending square tiling, but the length of each $^4/_2$ side in that tiling vertex is zero.

Truncating the crosspacks is where it gets interesting. The tetratile crosspack {4,4,$^4/_2$} truncates to produce what is essentially a series of octagon-square tilings perpendicular to each other, whilst the stellated hexahedral crosspack {$^4/_2$,4,4} produces a closepack of triple cubes extending to infinity. Beyond here further truncations complete a first series for the {4,4,$^4/_2$} family of crosspacks and we can also find the start of a second series and several odd crosspacks completely analogous to those found in the closepacks, all of which are shown in the pages following.

{4,4,$^4/_2$} FAMILY CROSSPACKS

{4,4,$^4/_2$}

ζ_4

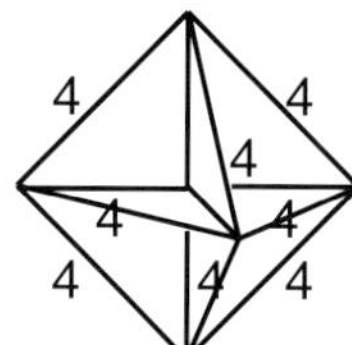

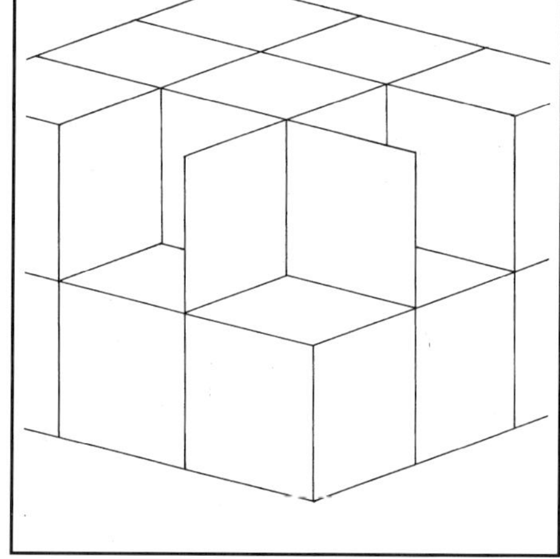

Tetratile Crosspack

t{4,4,$^4/_2$}

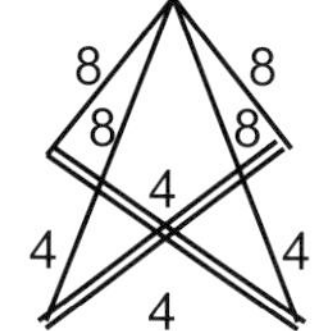

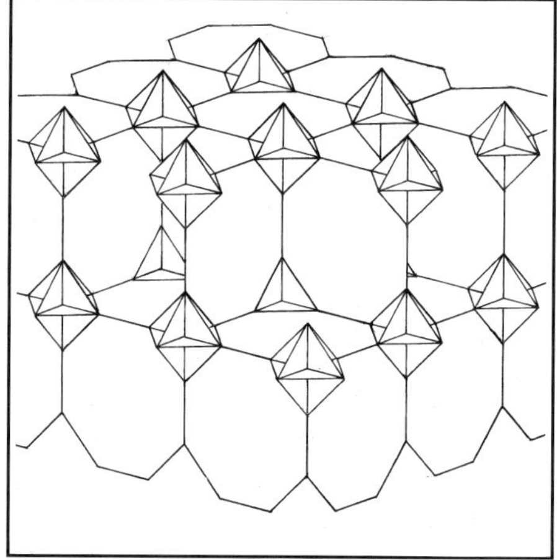

Truncated Tetratile Crosspack

t^2{4,4,$^4/_2$}

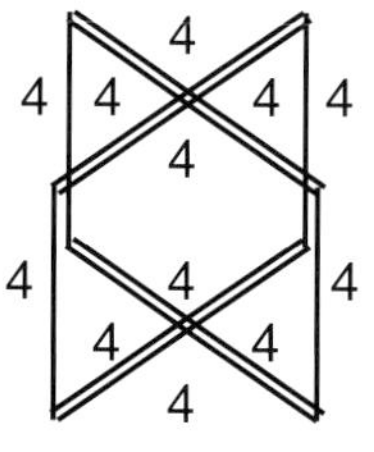

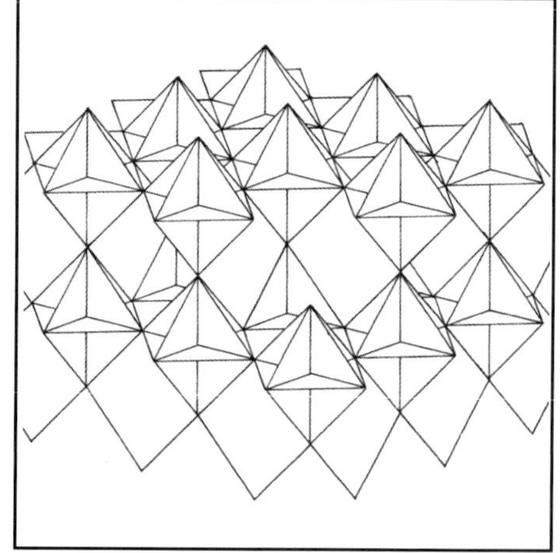

Hexatetrahedral Crosspack

$\{{}^4/_2,4,4\}$

ε_4

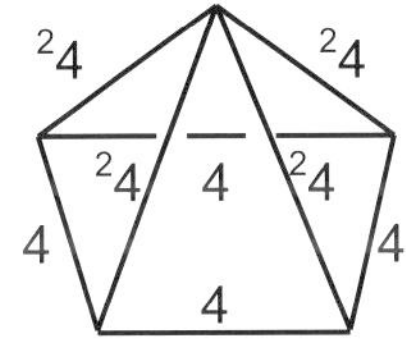

Stellated Hexahedral

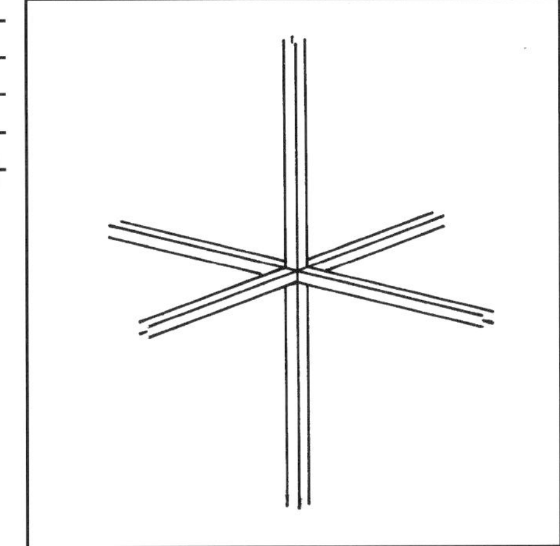

$t\{{}^4/_2,4,4\}$

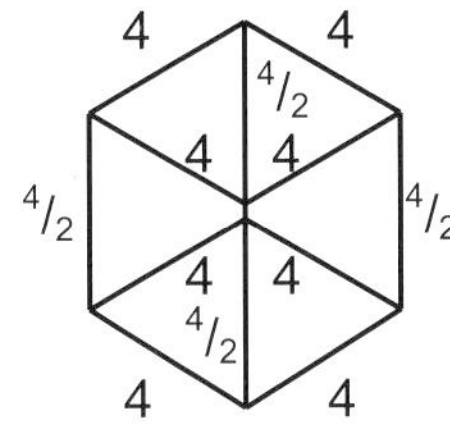

Truncated Stellated Hexahedral

Triple Cubic Closepack

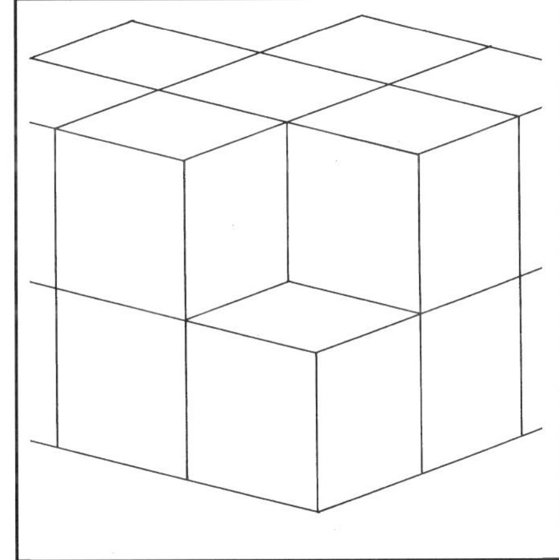

$t^2\{{}^4/_2,4,4\}$

Hexahexahedral Crosspack

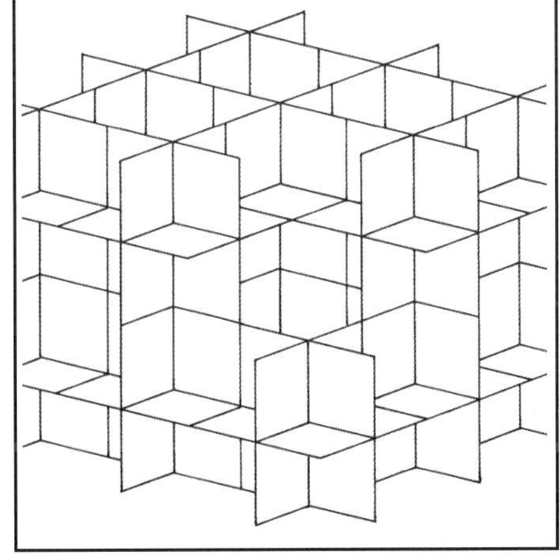

{4,4,^{4}/$_2$} FAMILY CROSSPACKS

t^3{4,4,^{4}/$_2$}

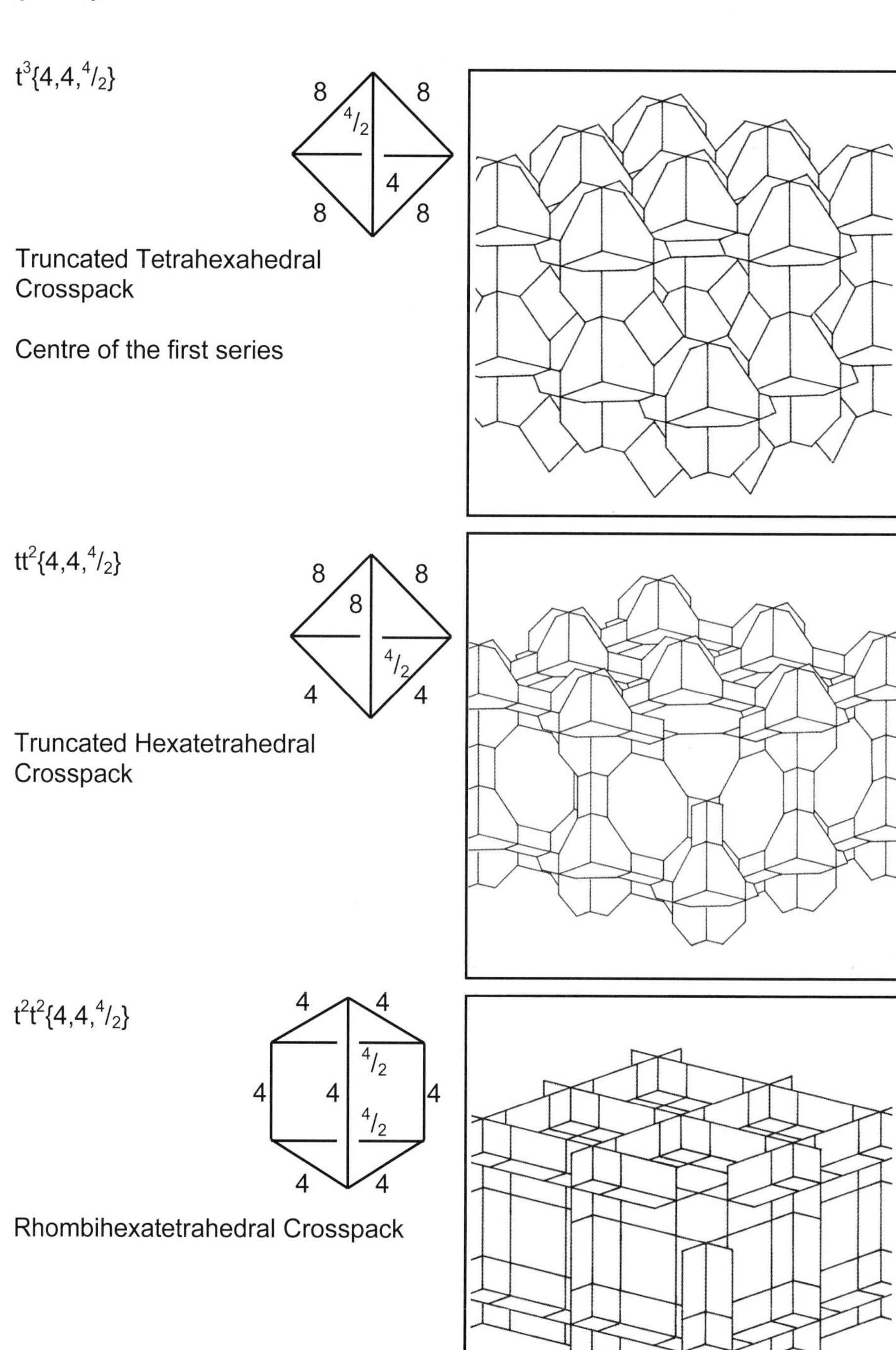

Truncated Tetrahexahedral
Crosspack

Centre of the first series

tt^2{4,4,^{4}/$_2$}

Truncated Hexatetrahedral
Crosspack

t^2t^2{4,4,^{4}/$_2$}

Rhombihexatetrahedral Crosspack

o^3{4,4,$^4/_2$}

Stellahexatetraprism Crosspack

tt^2{$^4/_2$,4,4}

Truncated Hexahexahedral
Crosspack

t^2t^2{$^4/_2$,4,4}

Rhombihexahexahedral Crosspack

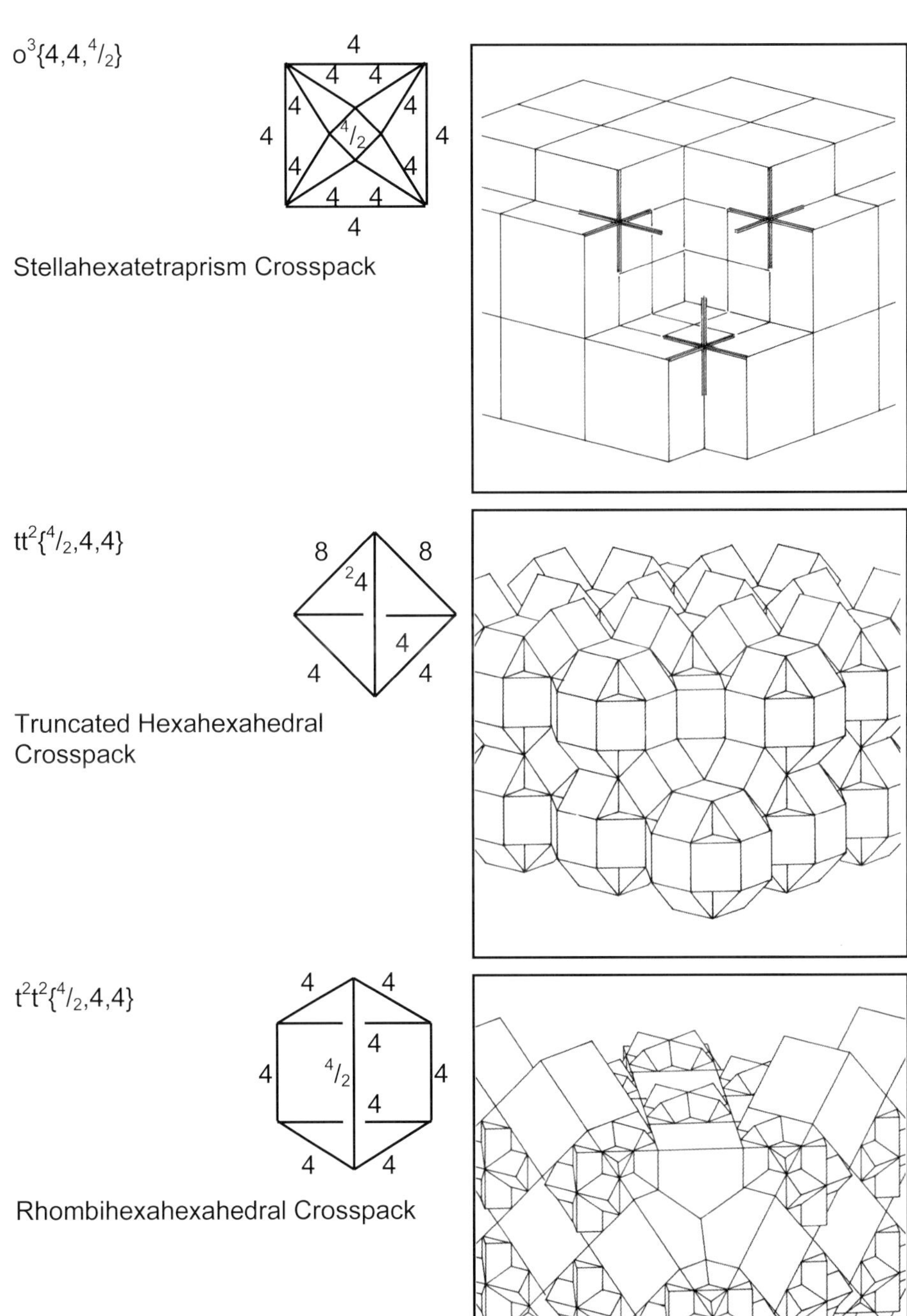

$\{4,4,{}^4/_2\}$ FAMILY CROSSPACKS

$o\{4,4,{}^4/_2\}$

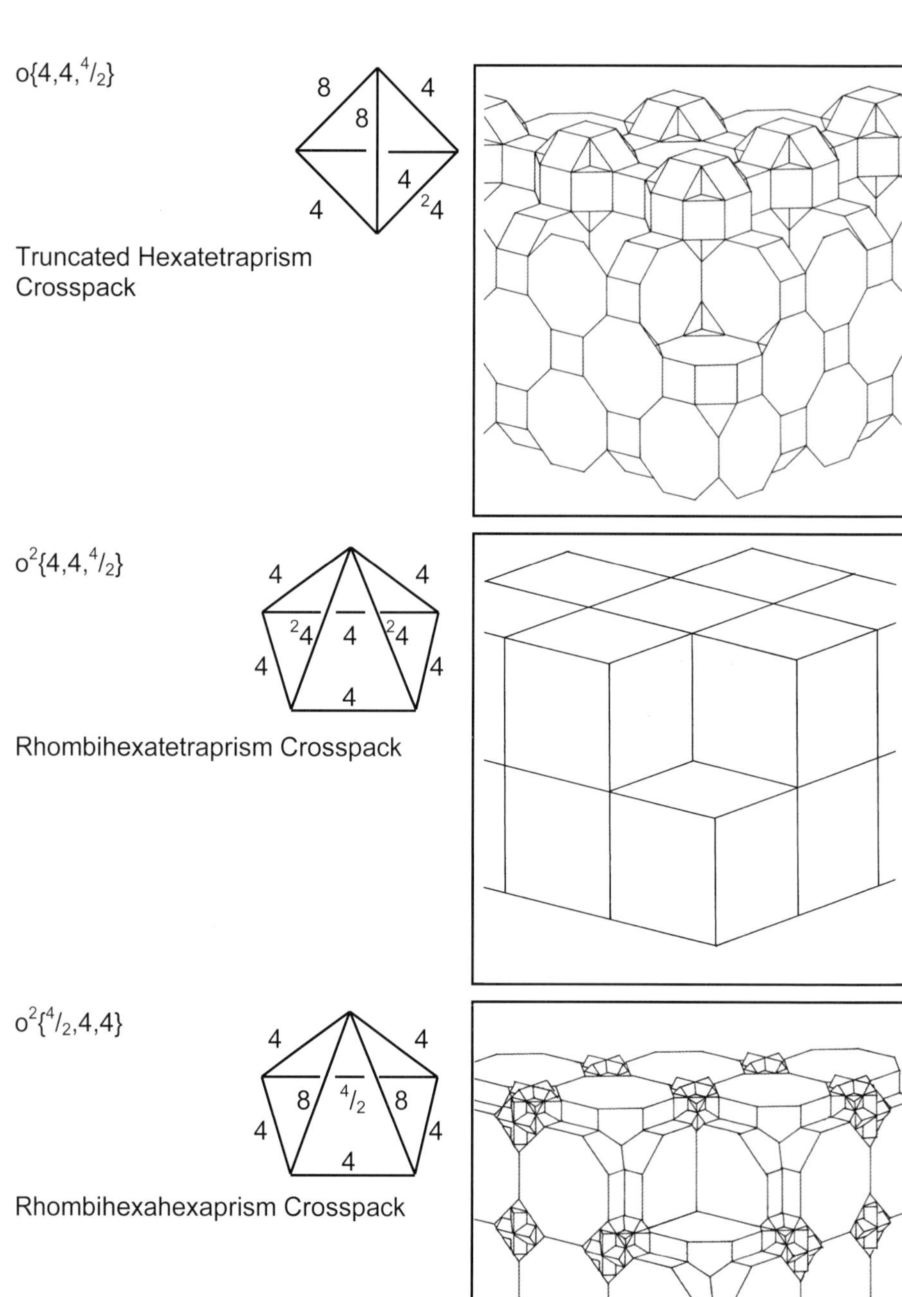

Truncated Hexatetraprism
Crosspack

$o^2\{4,4,{}^4/_2\}$

Rhombihexatetraprism Crosspack

$o^2\{{}^4/_2,4,4\}$

Rhombihexahexaprism Crosspack

The foregoing may not be particularly remarkable until one realises that the stellated hexahedral crosspack is indeed a strange beast. Despite the infinite number of stellated hexahedra it is composed of, it is surprisingly compact with zero volume, zero surface area and can only be measured linearly and finitely along any one of its three mutually perpendicular axes.

To go from this form, that you feel you can almost fold up and put in your pocket, to the infinitely extending closepack of triple cubes is indeed a big bang expressed geometrically. This is not a gradual expansion, for just clipping off the most minute portion at the tips of the six vertices leads the whole thing to suddenly fill three dimensions to infinity !

We will shortly see that this compact crosspack has analogues that repeat in higher dimensions with a general formula $\varepsilon_n = \{^4/_2,4,3^{n-4},4\}$. The analogue in four dimensions is ε_5 $\{^4/_2,4,3,4\}$ which has $\{^4/_2\}$ cross polygons arranged on an infinitely extending cubic closepack vertex figure. This has just the eight vertices of a stellated hypercube and similarly expands with a big bang to fill all of four dimensions immediately it is truncated.

CROSS HYPERSOLIDS

Apart from this 'big bang', the crosspacks appear to be well ordered and produce all the variations one would expect within a closepack symmetry family, complete with all the usual truncations and odd members with prism faces as illustrated here. The truncations shown all accord with those seen previously in the cross polyhedra and all obey the new truncation rules set out as part of the alternative notation used in this book.

The next chapter will look at the more compact forms produced by considering cross polyhedra assembled into finite collections folded up into four dimensions, direct analogues of the hypersolids themselves.

REFERENCES

Taylor, P. 1997 *The Complete? Polygon* Nattygrafix

Taylor, P. 1998 *Incomplete Tilings* Nattygrafix

Taylor, P. 1999 *The Simpler? Polyhedra* Nattygrafix

Taylor, P. 2000 *The Star & Cross Polyhedra* Nattygrafix

Taylor, P. 2004 *Closepacks & Quasi-Closepacks* Nattygrafix

Taylor, P. 2006 *Hypersolids & Quasi-Hypersolids* Nattygrafix

Taylor, P. 2007 *Crosspacks & Cross Hypersolids* Nattygrafix

In the same way that the hypersolids result from taking a closepack and putting fewer or smaller valued polyhedra around an edge, the cross hypersolids can be generated from the crosspacks.

With the closepacks we had the regular parent {4,3,4} and with hypersolids we had the regular parents {5,3,3}, {4,3,3}, {3,3,3}, {3,3,4}, {3,3,5} and {3,4,3}, all of which if taken as {x,y,z} had the property that {x,y} and {y,z} are both regular platonic solids, the former a constituent solid, the latter as a vertex figure.

Introducing the star polyhedra into the mix, we find the same rule applies with the only possible star hypersolids being the self-duals $\{^5/_2,5,^5/_2\}$ and $\{5,^5/_2,5\}$ along with the four dual pairs $\{^5/_2,5,3\}$ - $\{3,5,^5/_2\}$, $\{^5/_2,3,5\}$ - $\{5,3,^5/_2\}$, $\{5,^5/_2,3\}$ - $\{3,^5/_2,5\}$ and $\{^5/_2,3,3\}$ - $\{3,3,^5/_2\}$.

That same rule also applies to the new crosspacks we have just seen, where the cross polyhedra $\{4,^4/_2\}$ and $\{^4/_2,4\}$ can be used in combination with the square tiling {4,4} to produce the dual pair of crosspacks $\{4,4,^4/_2\}$ and $\{^4/_2,4,4\}$ as described in the last chapter.

However this rule will also potentially produce $\{3,4,^4/_2\}$, $\{^4/_2,4,3\}$ and $\{4,^4/_2,4\}$, the cross hypersolids. These comprise a dual pair and a self-dual and are illustrated along with their various truncations etc. in the sections that follow.

$\{3,4,^4/_2\}$

φ_4

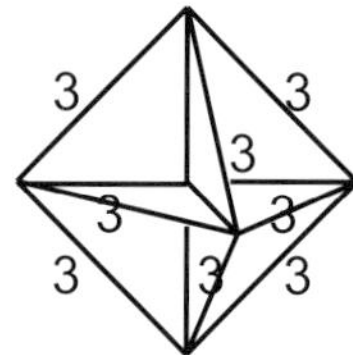

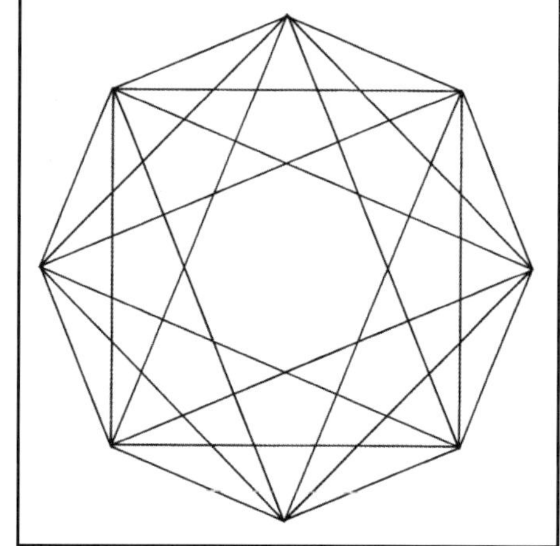

Octahyperhexahedron

$t\{3,4,^4/_2\}$

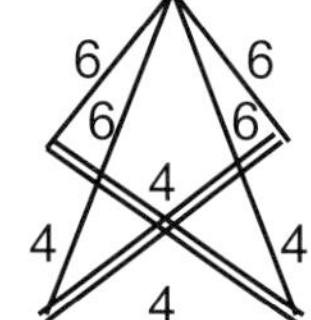

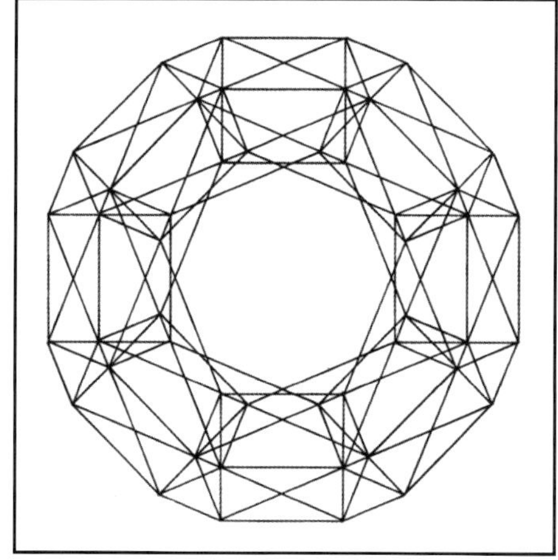

Truncated Octahyperhexahedron

$t^2\{3,4,^4/_2\}$

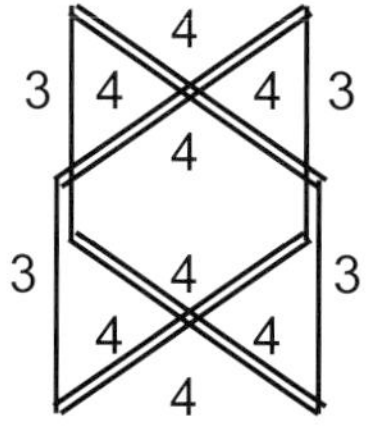

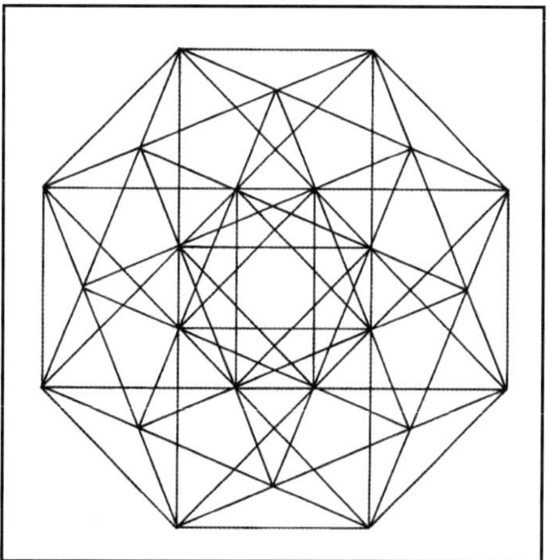

Great Hypercuboctahexahedron

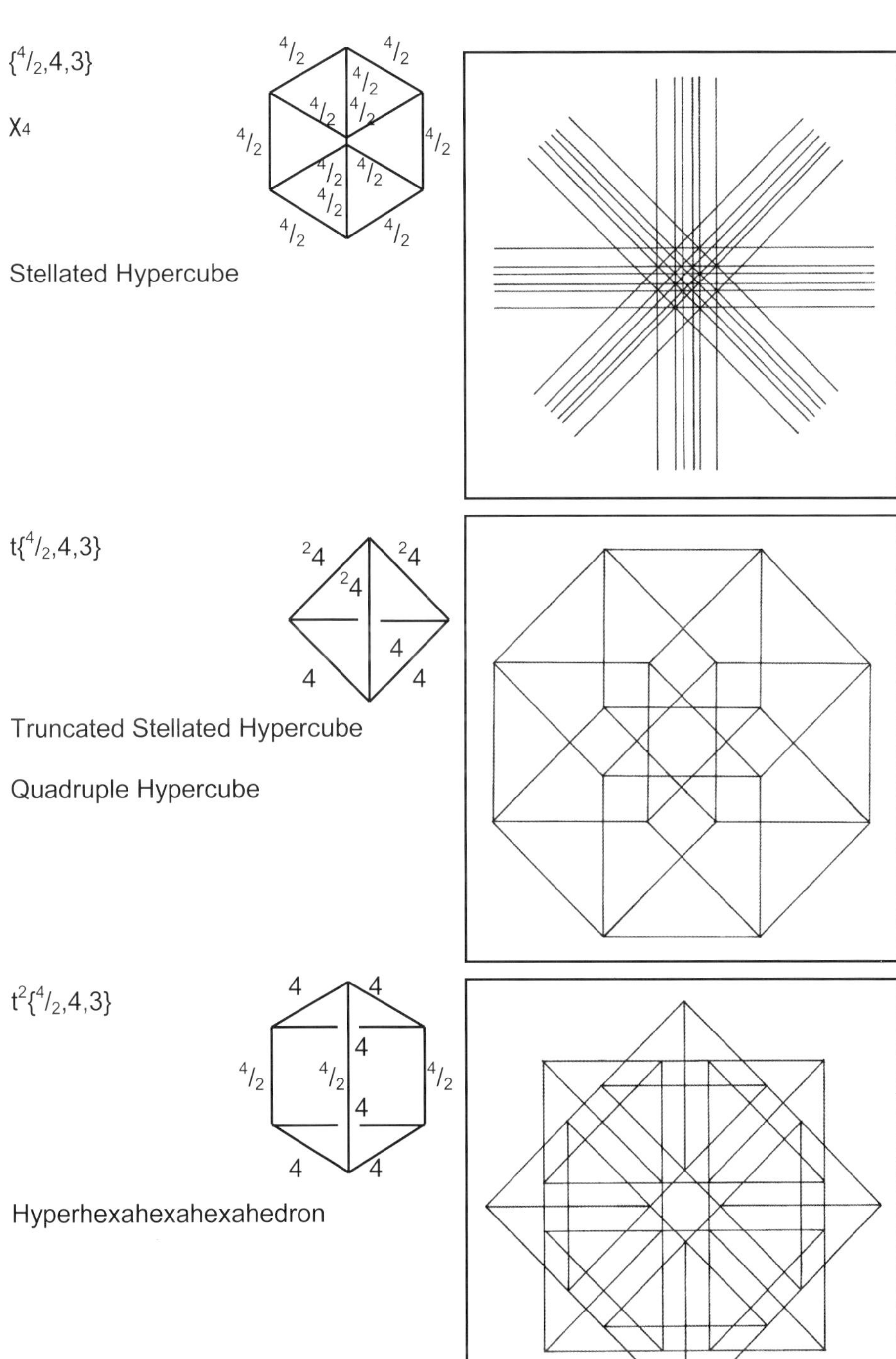

{$^4/_2$,4,3}

X$_4$

Stellated Hypercube

t{$^4/_2$,4,3}

Truncated Stellated Hypercube

Quadruple Hypercube

t^2{$^4/_2$,4,3}

Hyperhexahexahexahedron

{3,4,$^4/_2$} FAMILY CROSS HYPERSOLIDS

t^3{3,4,$^4/_2$}

Hypertruncated Hexahexahedron

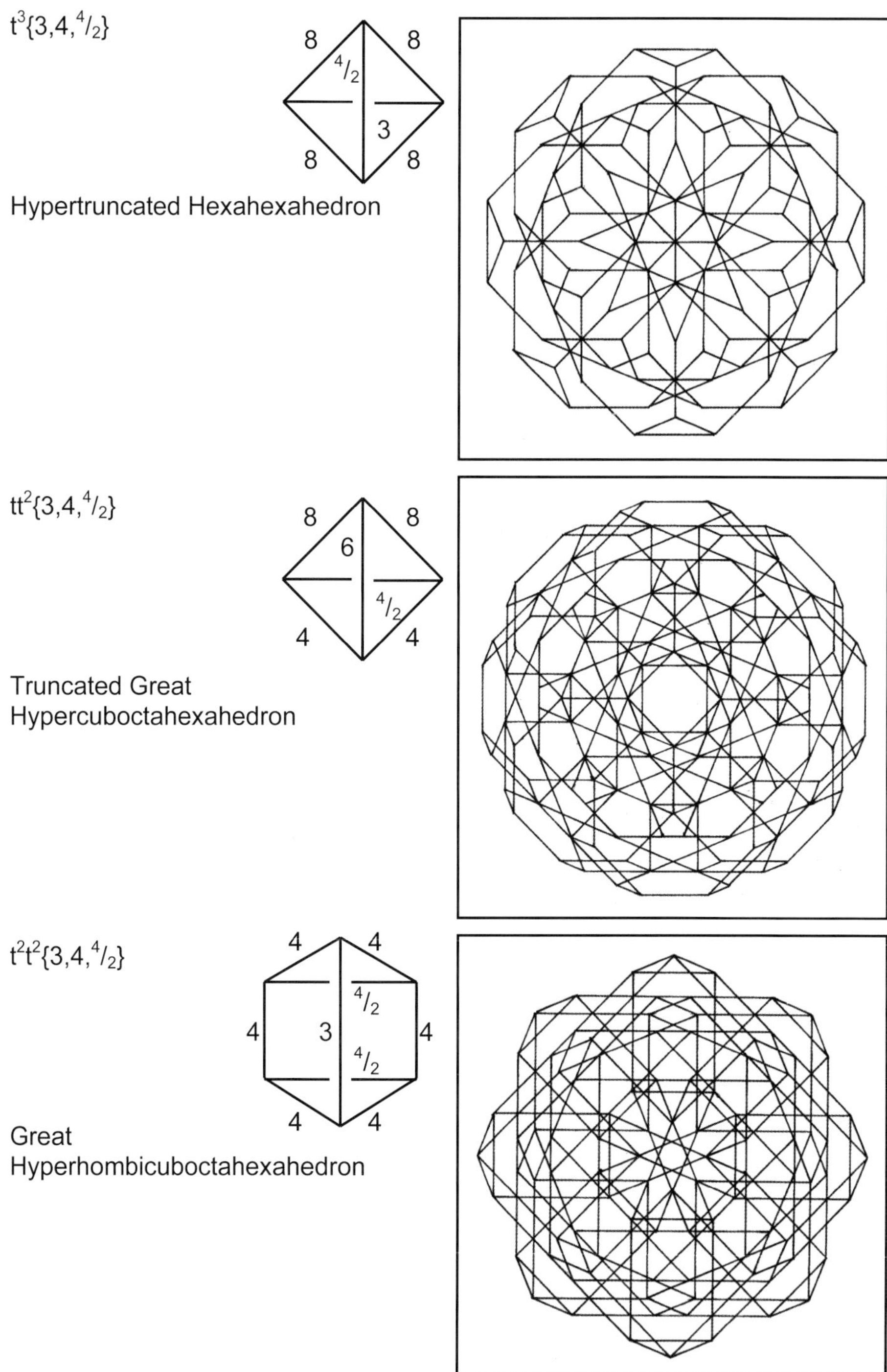

tt^2{3,4,$^4/_2$}

Truncated Great
Hypercuboctahexahedron

t^2t^2{3,4,$^4/_2$}

Great
Hyperhombicuboctahexahedron

$o^3\{3,4,{}^4/_2\}$

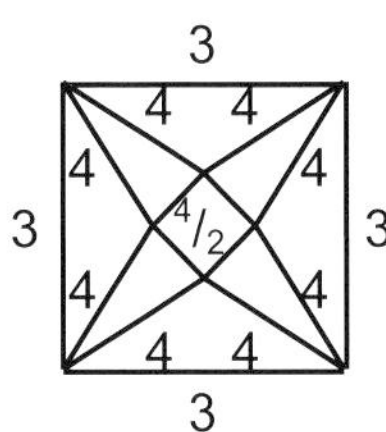

Stellated Hyperhexoctaprism

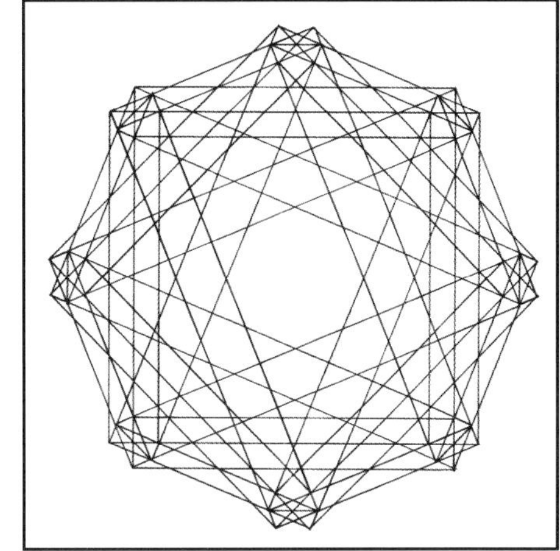

$tt^2\{{}^4/_2,4,3\}$

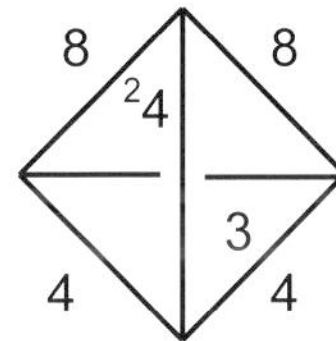

Truncated
Hyperhexahexahexahedron

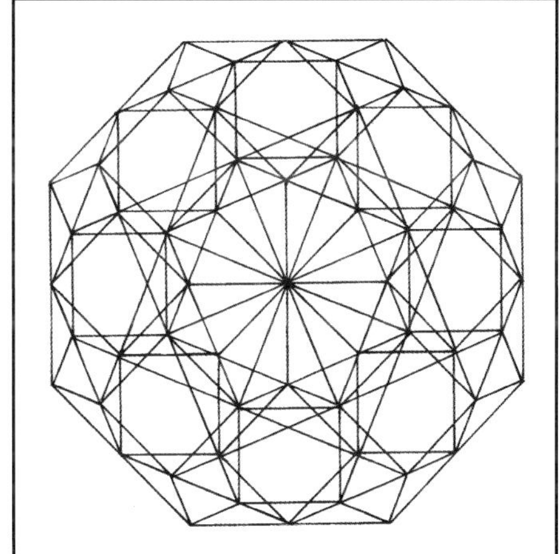

$t^2t^2\{{}^4/_2,4,3\}$

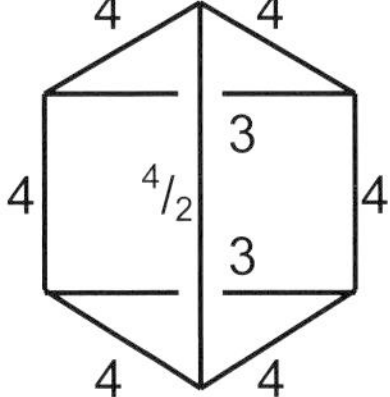

Hyperhombihexahexahexahedron

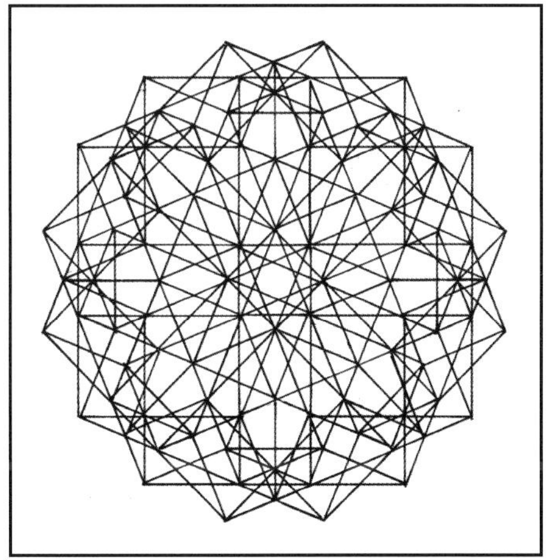

o{3,4,⁴/₂}

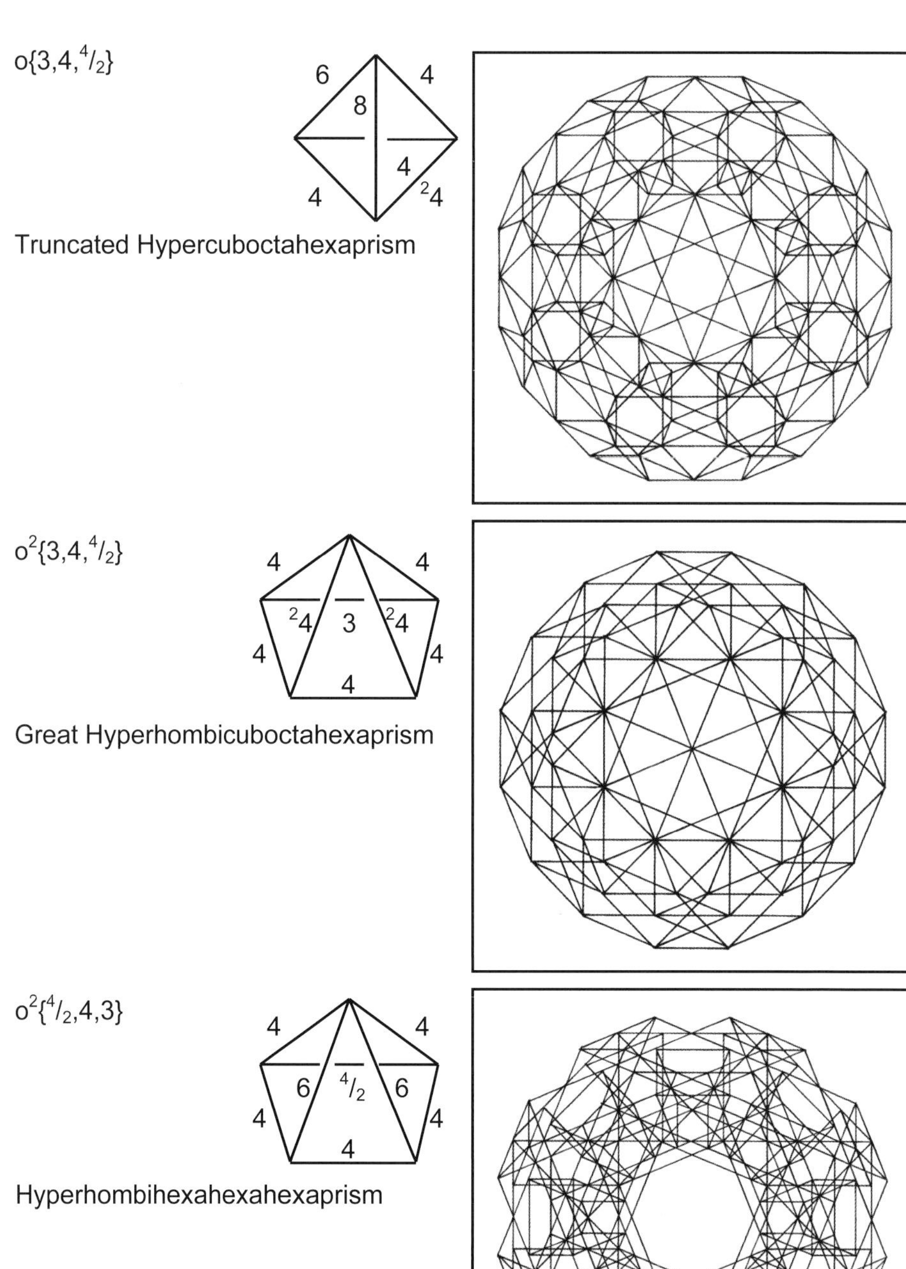

Truncated Hypercuboctahexaprism

o²{3,4,⁴/₂}

Great Hyperhombicuboctahexaprism

o²{⁴/₂,4,3}

Hyperhombihexahexahexaprism

{4,$^4/_2$,4}

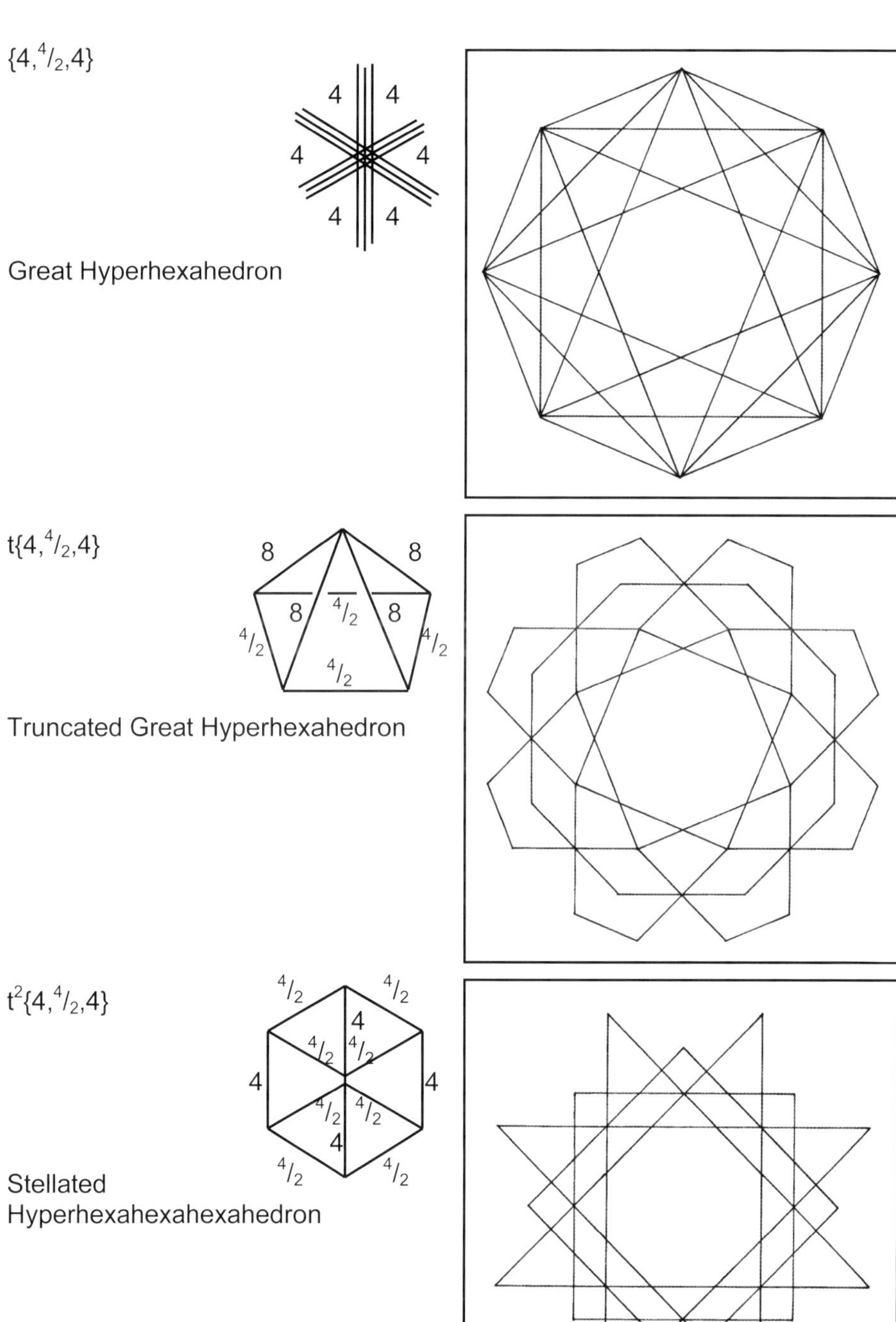

Great Hyperhexahedron

t{4,$^4/_2$,4}

Truncated Great Hyperhexahedron

t^2{4,$^4/_2$,4}

Stellated
Hyperhexahexahexahedron

{4,$^4/_2$,4} FAMILY CROSS HYPERSOLIDS

$t^3\{4,^4/_2,4\}$

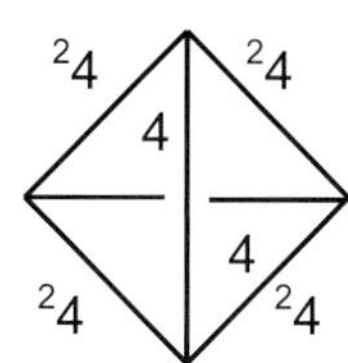

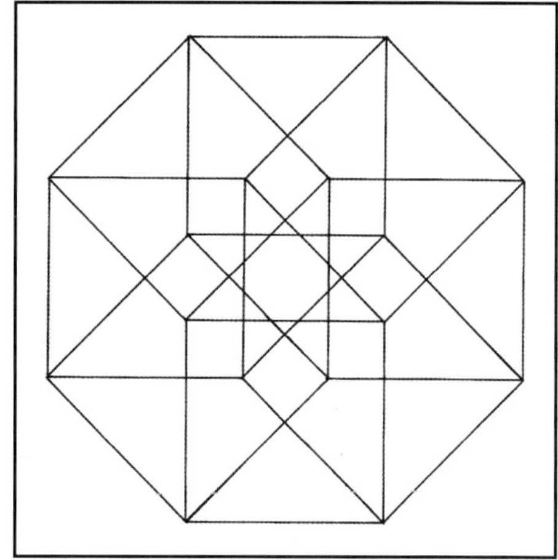

Stellated Hypertruncated
Hexahexahedron

Sixfold Hypercube

$tt^2\{4,^4/_2,4\}$

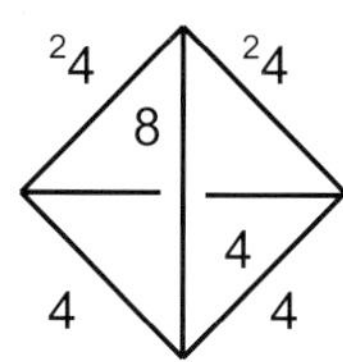

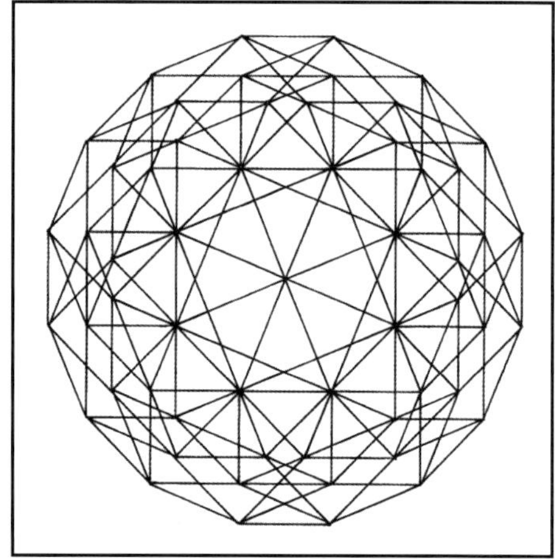

Truncated Stellated
Hyperhexahexahexahedron

$t^2t^2\{4,^4/_2,4\}$

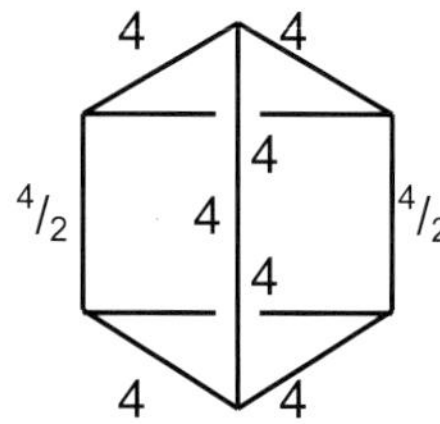

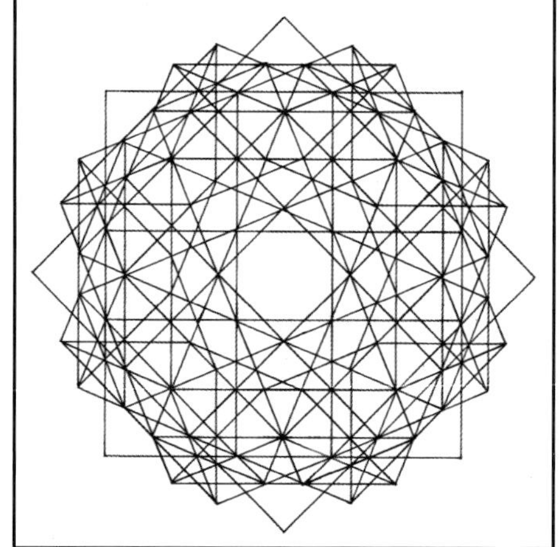

Stellated
Hyperhombihexahexahedron

$o\{4,{}^4/_2,4\}$

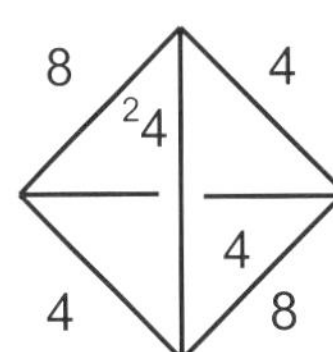

**Truncated Stellated
Hyperhexahexahexaprism**

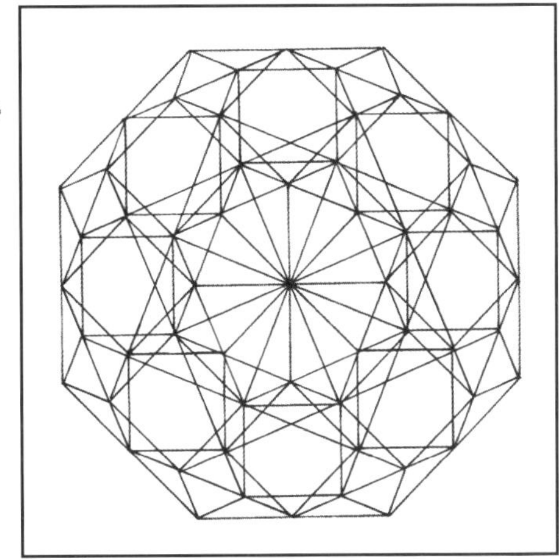

$o^2\{4,{}^4/_2,4\}$

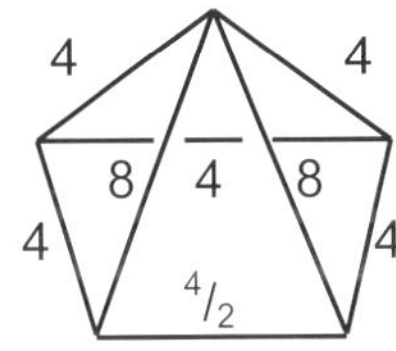

**Stellated
Hyperhombihexahexahexaprism**

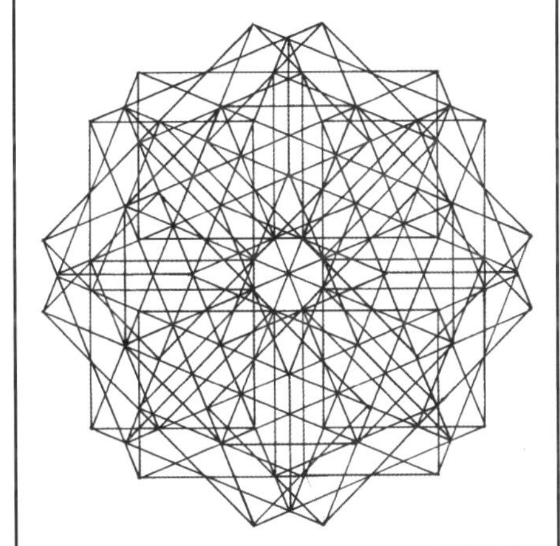

$o^3\{4,{}^4/_2,4\}$

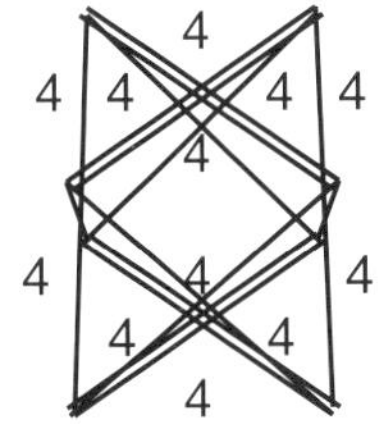

Great Hyperhexahexaprism

The vertex figure for this cross
hypersolid is a $\{^4/_2\}$ cross polygon
antiprism, a stella octangula

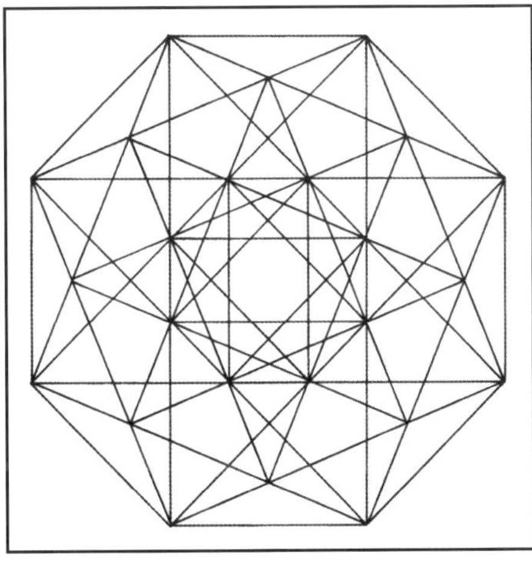

POLYTOPES

We have already seen that a number of regular polytopes as described on p.89 repeat as we go up through the dimensions with standard formulae:

Simplex $\quad\quad\quad\quad \alpha_n = \{3^{n-1}\} \quad\quad\quad$ Dual $\quad\quad\quad\quad \beta_n = \{3^{n-2},4\}$
Measure $\quad\quad\quad \gamma_n = \{4,3^{n-2}\} \quad\quad$ Honeycomb $\quad\quad \delta_n = \{4,3^{n-3},4\}$

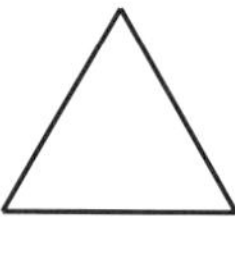 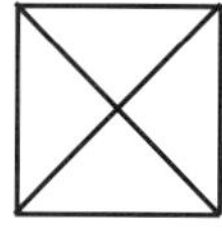 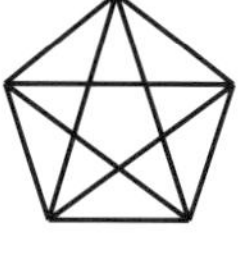 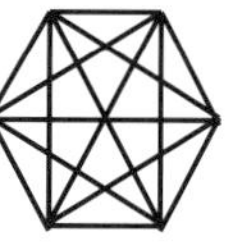

$\alpha_2 = \{3\} \quad\quad \alpha_3 = \{3,3\} \quad\quad \alpha_4 = \{3,3,3\} \quad \alpha_5 = \{3,3,3,3\}$ etc.

The simplex polytope in n dimensions will have n+1 vertices equally spaced away from a common origin and can be represented in two dimensions as an (n+1)gon with every vertex connected to every other.

 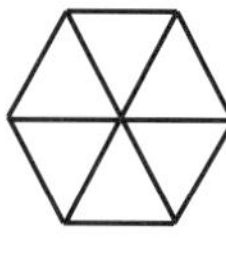 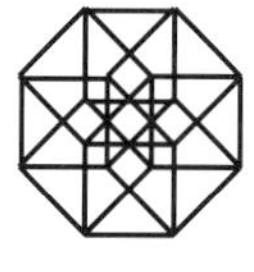 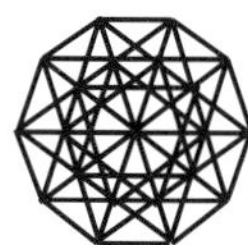

$\gamma_2 = \{4\} \quad\quad \gamma_3 = \{4,3\} \quad\quad \gamma_4 = \{4,3,3\} \quad \gamma_5 = \{4,3,3,3\}$ etc.

The measure polytope in n dimensions will have 2^n vertices and can be represented in two dimensions as a (2n)gon with (2^n-2n) extra internal vertices. Its 2n cells (edges, faces, solids etc.) have centres that form the vertices of its dual, all equidistant from the origin of an n-dimensional Cartesian frame of reference.

 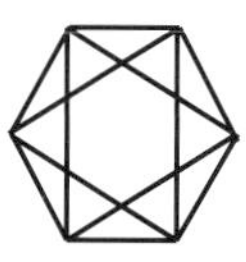 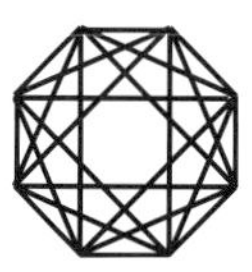 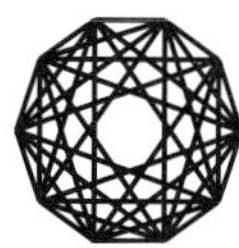

$\beta_2 = \{4\} \quad\quad \beta_3 = \{3,4\} \quad\quad \beta_4 = \{3,3,4\} \quad \beta_5 = \{3,3,3,4\}$ etc.

The dual polytope in n dimensions will thus have 2n vertices occurring in opposing pairs and can be represented in two dimensions as a (2n)gon with every vertex connected to every other except the one opposite. It is the disposition of these vertices that has led to these being named in the standard version as 'cross polytopes'. However as we have seen these vertices also belong to further polytopes that better deserve the name, the true cross polytopes.

The crosspacks and cross hypersolids can similarly be seen as a number of cross polytopes that repeat at each dimension with standard formulae:

Dualcross $\quad\quad \varphi_n = \{3^{n-3},4,{}^4/_2\} \quad$ Cross $\quad\quad \chi_n = \{{}^4/_2,4,3^{n-3}\}$

Crosspack $\quad\quad \varepsilon_n = \{{}^4/_2,4,3^{n-4},4\} \quad$ Dualpack $\quad\quad \zeta_n = \{4,3^{n-4},4,{}^4/_2\}$

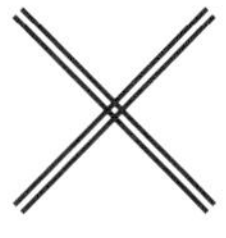 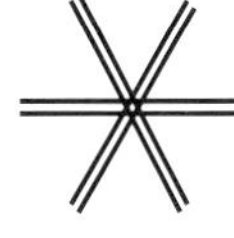 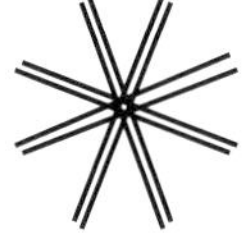 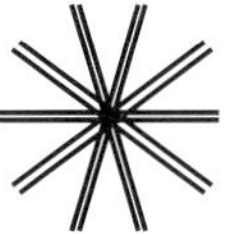

$$\chi_2 = \{{}^4/_2\} \quad\quad \chi_3 = \{{}^4/_2,4\} \quad\quad \chi_4 = \{{}^4/_2,4,3\} \quad\quad \chi_5 = \{{}^4/_2,4,3,3\} \text{ etc.}$$

Sticking to our alternative notation, the true cross polytope in n dimensions will have 2n vertices the same as those of the dual polytope. It can be represented in two dimensions simply by the axes of an n dimensional Cartesian frame, just the missing diagonals of the dual polytope's representation, a true n-dimensional 'cross'.

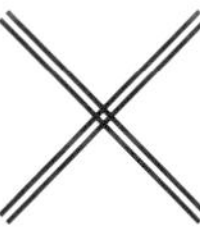 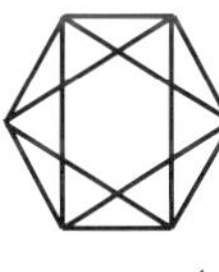 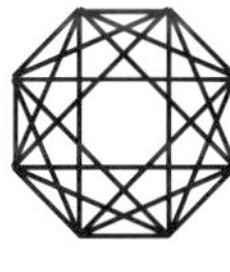 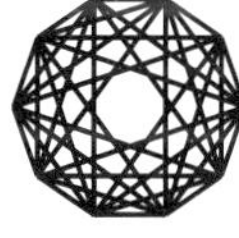

$$\varphi_2 = \{{}^4/_2\} \quad\quad \varphi_3 = \{4,{}^4/_2\} \quad\quad \varphi_4 = \{3,4,{}^4/_2\} \quad\quad \varphi_5 = \{3,3,4,{}^4/_2\} \text{ etc.}$$

Its dual, the dualcross polytope in n dimensions will also have 2n vertices and its representation in two dimensions will generally be the same as that of the corresponding dual polytope, in this case showing exactly the same edges, but with different faces inscribed between.

Strictly speaking the cross and dualcross polytope formulae $\chi_n = \{{}^4/_2,4,3^{n-3}\}$ and $\varphi_n = \{3^{n-3},4,{}^4/_2\}$ do not apply when n = 2, as the formulae at higher dimensions simply add or subtract $\{3\}$ faces. The diagrams show χ_2 as a cross polygon $\{{}^4/_2\}$ which fits well, whilst φ_2 should perhaps be a square $\{4\}$, but this polytope is actually inscribed within a square and has to be the dual of χ_2 so correctly it is a $\{{}^4/_2\}$ too.

STELLATION & INSCRIPTION

The cross polytopes can be linked with the regular polytopes by a system of substitution. This works by considering $\beta_n = \{3^{n-3},3,4\}$, $\gamma_n = \{4,3,3^{n-3}\}$ and $\delta_n = \{4,3,3^{n-4},4\}$ or $\{4,3^{n-4},3,4\}$ and then substituting $\{^4/_2,4\}$ for $\{4,3\}$ and $\{4,^4/_2\}$ for $\{3,4\}$.

Now the $\{^4/_2\}$ cross polygon was seen at the outset as either a stellation or an inscription of a square $\{4\}$. These substitutions that will change regular polytopes into cross polytopes are thus effectively the stellation of a cube where $\{4,3\}$ becomes $\{^4/_2,4\}$ and the inscription an octahedron where $\{3,4\}$ becomes $\{4,^4/_2\}$. An encouraging symmetry!

In general we thus find that the cross polytope $\chi_n = \{^4/_2,4,3^{n-3}\}$ is simply a stellated measure polytope $\gamma_n = \{4,3^{n-2}\}$ and its dual the dualcross polytope $\varphi_n = \{3^{n-3},4,^4/_2\}$ is simply an inscribed dual polytope $\beta_n = \{3^{n-2},4\}$.

A parallel relationship holds for the star polyhedron $\{^5/_2,5\}$ being a stellation of the dodecahedron $\{5,3\}$ and its dual the great dodecahedron $\{5,^5/_2\}$ being an inscription within an icosahedron $\{3,5\}$. A dimension higher we can see the star hypersolid $\{^5/_2,5,3\}$ as a stellated hyperdodecahedron $\{5,3,3\}$ and its dual $\{3,5,^5/_2\}$ as an inscribed icosahypertetrahedron $\{3,3,5\}$.

Turning next to the closepacks, we can see that similarly the crosspack and dualpack polytopes $\varepsilon_n = \{^4/_2,4,3^{n-4},4\}$ and $\zeta_n = \{4,3^{n-4},4,^4/_2\}$ are nothing other than stellated and inscribed versions respectively of the honeycomb polytope $\delta_n = \{4,3^{n-3},4\}$.

HIGHER DIMENSIONS

Moving up a dimension from the hypersolids, we already know that there will be a simplex polytope α_5 $\{3,3,3,3\}$, a measure polytope γ_5 $\{4,3,3,3\}$ and a dual polytope β_5 $\{3,3,3,4\}$ that form finite assemblies in 5-space.

Using our method of substitution at this level leads us to a five dimensional cross polytope $\chi_5 = \{^4/_2,4,3,3\}$ plus a dualcross polytope $\varphi_5 = \{3,3,4,^4/_2\}$.

Filling four dimensions there are also three polytopes known as 'honeycombs', a general term used to describe both the tilings and closepacks in the lower dimensions.

The general case of square tiling {4,4} δ_3 in two dimensions and cubic closepack {4,3,4} δ_4 in three has a parallel δ_5 in four dimensions given in the Schläfli notation as {4,3,3,4}. Like the lower members it is self dual and can be considered either as hypercubes {4,3,3} arranged four per edge or squares {4} arranged around an octahypertetrahedral {3,3,4} vertex figure.

The cross honeycombs that result from the two possible substitutions into δ_5 {4,3,3,4} are ϵ_5 {$^4/_2$,4,3,4} and ζ_5 {4,3,4,$^4/_2$}. If both substitutions are carried out there may well also be a viable honeycomb given as {$^4/_2$,4,4,$^4/_2$}.

Back with the regular honeycombs there is also a dual pair, given as {3,4,3,3} and {3,3,4,3}.   The former can be considered as either hyperoctahedra {3,4,3} arranged three per edge or triangles {3} arranged around a hypercubic {4,3,3} vertex figure.   The latter can likewise be imagined as either octahypertetrahedra {3,3,4} arranged three per edge or triangles {3} arranged around a hyperoctahedral {3,4,3} vertex figure.

If, as seems likely, there are further cross honeycombs at this level of organisation filling four dimensions, they will potentially comprise three dual pairs. These are with varying levels of substitution {4,$^4/_2$,3,3} and {3,3,$^4/_2$,4}, {3,$^4/_2$,4,3} and {3,4,$^4/_2$,3} and finally {4,$^4/_2$,4,3} and {3,4,$^4/_2$,4}.

Difficulty of illustration means that it is beyond the scope of this book to go any further with this, so these forms must remain predictions that will hopefully be borne out by those more atuned to working in higher dimensions, probably using high powered computers.

REFERENCES

Taylor, P. 1997 *The Complete? Polygon* Nattygrafix

Taylor, P. 1998 *Incomplete Tilings* Nattygrafix

Taylor, P. 1999 *The Simpler? Polyhedra* Nattygrafix

Taylor, P. 2000 *The Star & Cross Polyhedra* Nattygrafix

Taylor, P. 2004 *Closepacks & Quasi-Closepacks* Nattygrafix

Taylor, P. 2006 *Hypersolids & Quasi-Hypersolids* Nattygrafix

Taylor, P. 2007 *Crosspacks & Cross Hypersolids* Nattygrafix

So there we have it, the simple tweaking of the definition of a polygon using an alternative notation has led firstly to a consistent system of naming the compound star and cross polygons. New families of star and cross tilings followed along with the cross polyhedra, sometimes called the Taylor Polyhedra, originally described in the first paper of the author's *Additions to the Uniform Polyhedra*.

The second paper of *Additions to the Uniform Polyhedra* dealt with polyhedra with ditrigonal vertex figures, which result naturally from a proper consideration of the double polygons that arise when an even denominator polygon is truncated. Like the currently accepted notation, the alternative notation also led to the production of double coincident polygons when even denominator polygons are truncated.

These are the only multiple polygons that need to be considered when dealing with tilings, polyhedra and the higher dimension polytopes. The triple, quadruple and higher multiple coincident polygons produced by the currently accepted notation are effectively spurious and of no further use. The only place for higher multiples is when they are out of phase and as described by the alternative notation. The new cross polyhedra in their turn have now been assembled into the higher dimensional examples of crosspacks (filling three dimensions like the closepacks) and the four dimensional cross hypersolids which share their edges with various regular hypersolids.

Looking at the subject in a more general way, we have just seen that the usual measure (and dual) and honeycomb polytopes all have parallels in the cross polytopes, the latter being simply derived from the former by two simple substitutions in the general formulae that equate to the processes of stellation or inscription.

In an age when we hear so often how the big questions of cosmology and particle physics are intimately bound up with matters of symmetry, surely we need to have our interpretation of geometry admitting of the full range of possibilities, rather than restricted to a narrow interpretation linked to number theory, as used by the currently accepted notation for polygons and whatever we might construct with them.

'If we do it this way, then what follows?' is not an incorrect way of doing mathematics, just a different way using an alternative notation, which is hopefully now proven worthwhile by the results shown herein. Of course now we have seen the results, all this could be done using the currently accepted notation for polygons, the results hold whatever, but their description using that system would be somewhat tedious, clumsy and certainly less than obvious.

COLLECTED REFERENCES

Coxeter, H.S.M. 1973 *Regular Polytopes* Dover, New York

Coxeter, H.S.M., Longuet-Higgins, M.S. & Miller, J.C.P. 1953 *Uniform Polyhedra* Phil. Trans. R. Soc. Lond. A 246, 401-449

Critchlow, K. 1969 *Order in Space* Thames & Hudson

Cundy, H.M. & Rollett, A.P. 1951 *Mathematical Models* Oxford U P

Escher, M.C. 1992 *The Graphic Work* Benedikt Taschen

Grünbaum, B. & Shephard, G.C. 1989 *Tilings and Patterns; An Introduction* W.H.Freeman

Grünbaum, B. 2003 *Are Your Polyhedra the Same as My Polyhedra?* In: Aronov, Basu, Pach & Sharir (eds) Discrete and Computational Geometry. Algorithms and Combinatorics, vol 25 Springer, Berlin, Heidelberg

Holden, A. 1971 *Shapes, Space, and Symmetry* Columbia U P

Johnston, J.H. 1937 *The Reverse Notation* Blackie & Son Ltd

Kasparian, R.J. & Petillo, A.E. 2017 *Introducing the Kasparian Constructions* Bridges 2017 Conference Proceedings

Open University 1994 *Tilings* M336 Block 1, Unit IB1

Skilling, J. 1974 *The Complete Set of Uniform Polyhedra* Phil. Trans. R. Soc. Lond. A 278, 111-135

Taylor, P. 1995 *Additions to the Uniform Polyhedra* Nattygrafix

Taylor, P. 1997 *The Complete? Polygon* Nattygrafix

Taylor, P. 1998 *Incomplete Tilings* Nattygrafix

Taylor, P. 1999 *The Simpler? Polyhedra* Nattygrafix

Taylor, P. 2000 *The Star & Cross Polyhedra* Nattygrafix

Taylor, P. 2004 *Closepacks & Quasi-Closepacks* Nattygrafix

Taylor, P. 2006 *Hypersolids & Quasi-Hypersolids* Nattygrafix

Taylor, P. 2007 *Crosspacks & Cross Hypersolids* Nattygrafix

Wenninger, M.J. 1971 *Polyhedron Models* Cambridge U P

Wenninger, M.J. 1983 *Dual Models* Cambridge U P

HYPERBLOT: A DIVERSION

It is sometimes said that the outline of the Mandelbrot set, M, is the most complex shape known to man. In the two dimensions of the complex plane $(z = x + iy : i^2 = -1)$ it represents the set of starting points $\{z_1\}$ which when repeatedly iterated $z_{n+1} = z_n^2 + z_1$, do not tend to infinity.

In a similar way that two dimensional geometry can be represented by operations in the complex plane, our four dimensional world of hyperspace can be represented by a system of 'quaternions': $V = (w) + (ix + jy + kz)$.

In this system a hypercube of edge length 2 centred at the origin will have sixteen (2^4) vertices represented by $(\pm 1) + (\pm i \pm j \pm k)$, much the same as a square's vertices in the complex plane, which are $(\pm 1) + (\pm i)$.

In general quaternions consist of a scalar part w and a vector part $v = ix + jy + kz$ such that $i^2 = j^2 = k^2 = -1$ and $ij = k$, $jk = i$ and $ki = j$.

The rule for 'multiplying' two quaternions is $V_1V_2 = (w_1w_2 - v_1.v_2) + (w_1v_2 + w_2v_1 + v_1{}_xv_2)$, where $v_1.v_2$ and $v_1{}_xv_2$ are the vector dot product (a scalar) and vector cross product (a vector) respectively.

Thus to square a quaternion we get $VV = (ww - v.v) + (wv + wv + v_xv)$, or more simply $V^2 = (w^2 - v.v) + (2wv + v_xv)$.   But $v.v = |v|^2 = x^2 + y^2 + z^2$ and $v_xv = 0$, so $V^2 = (w^2 - x^2 - y^2 - z^2) + 2w(xi + yj + zk)$.

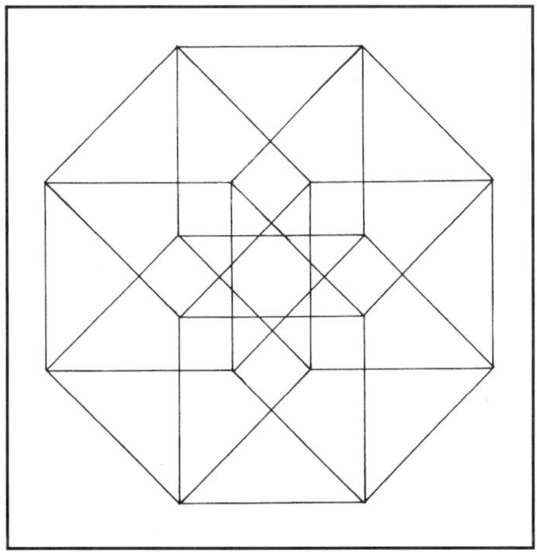

Hypercube

ITERATION

If we use this method of squaring as a basis for the iteration of quaternions $V_{n+1} = V_n^2 + V_1$, an interesting form emerges containing the set of points $\{V_1\}$ which do not go off to infinity.

The difficulty here is visualising this beast which now arises. With the hypercube mentioned above, setting any pair of w, x, y or z to zero leaves us simply with a square, a two dimensional section. Similarly setting just one of w, x, y or z to zero produces a cube as a sort of three dimensional section.

So with our 'hyperblot' setting any pair of our directions to zero chosen from x, y or z leaves us with one imaginary and one real dimension (e.g. x = y = 0 : V = (w) + (kz)). The iteration stays within the remaining plane and produces M, the Mandelbrot set, as a two dimensional section.

If the pair eliminated includes our real axis (e.g. w = z = 0 : V = (ix + jy)), the iteration straight away adds back in a real value w. The two dimensional section here is a series of concentric circles, infinite but necessarily odd in number, as we end up outside the set eventually. These circles are centred at the origin and range in radius from $\frac{1}{4}\sqrt{(3 + 2\sqrt{3})}$ where the basic cardioid of M is crossed by the imaginary axis, up to a final value near 1 where the iteration escapes to infinity.

If we similarly try for a three dimensional section, setting any one of x, y or z to zero, the iteration stays in the same three-space and produces what is essentially a solid of rotation of the Mandelbrot set symmetrically about the real axis.

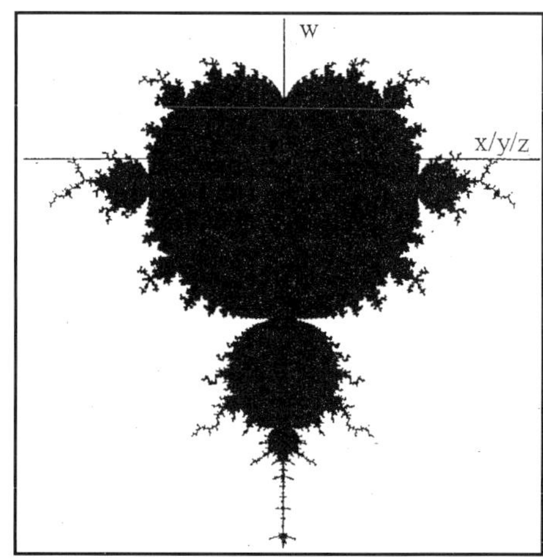

The Mandelbrot Set as Section

If we set w to zero and try to eliminate the real axis, again the iteration reintroduces it. Here the three dimensional 'section' is a set of concentric spheres of the same radii as the circles encountered above. Starting at the origin, again one first leaves the set at radius $\frac{1}{4}\sqrt{(3 + 2\sqrt{3})}$ and then one goes in and out of the set an infinite number of times before finally entering the area of escaping iteration.

So we seem to have here with our 'hyperblot' a four dimensional analogue of the Mandelbrot set. Three out of six of its two dimensional sections present us with M, the Mandelbrot set itself, whilst three out of four of its three dimensional sections are the solid of rotation of M about the real axis.

It is impossible to illustrate such a beast without the aid of complex computer graphics, so the author will leave this as a challenge to a brighter spark. However to help visualise this Hyperblot in some way, the figure facing shows the Mandelbrot set, albeit unusually with the real axis vertical. This shows fairly minimal detail; but the problem here is that the closer you look at the boundary the more you find, detail is laid on detail on detail to infinity.

A question now arises: this 'hyperblot' will undoubtedly represent a set of co-ordinates where the corresponding four dimensional quaternion 'Julia' sets are connected and their enclosed 'Keep' sets include 0. Their two dimensional sections have rotational symmetry, so what shape are they in four?

REFERENCES

Conway, J.H. & Smith, D.A. 2003 *On Quaternions and Octonians* AK Peters

Mandelbrot, B.B. 1983 *The Fractal Geometry of Nature* WH Freeman

Taylor, P. 2006 *Hypersolids & Quasi-Hypersolids* Nattygrafix

The Star & Cross Polyhedra

Patrick Taylor   2000   £6.00
ISBN 0 9516701 5 8   Nattygrafix

The fourth part of several comprising 'The Complete? Polyhedra'

presents the more complex polyhedra with star and cross polygon faces; their truncations, quasitruncations, snub forms plus sub-families and hemihedra, along with an appendix describing the prismatic star dihedral forms.

80pp A4 paperback including 71 line drawings and 37 tables

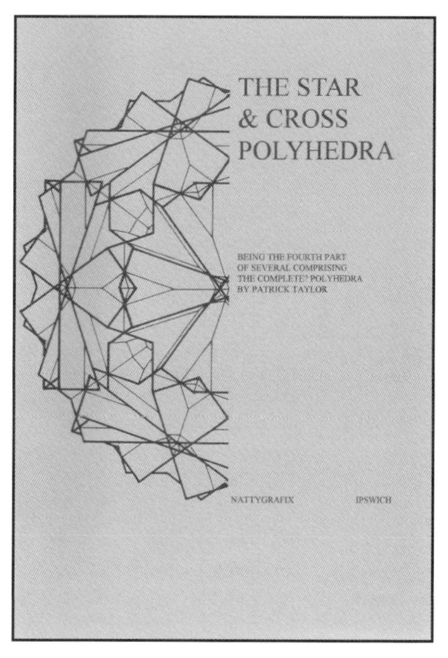

Crosspacks & Cross Hypersolids

Patrick Taylor   2007   £6.00
ISBN 0 9516701 8 2   Nattygrafix

The seventh part of several comprising 'The Complete? Polyhedra'

This volume takes the cross polyhedra that incorporate the $\{^4/_2\}$ cross polygon a step further into three and four dimensions where equivalents are found to the closepacks and hypersolids.

80pp A4 paperback including 70 line drawings and 55 tables

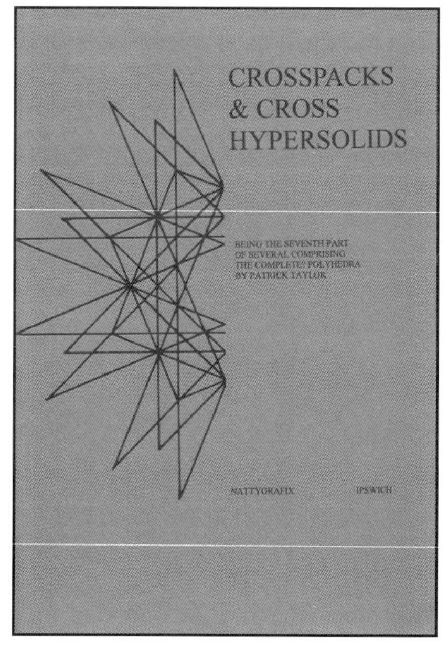